LIGHTS ON!

LIGHTS ON!
THE SCIENCE OF POWER GENERATION

MARK DENNY

THE JOHNS HOPKINS UNIVERSITY PRESS

BALTIMORE

© 2013 The Johns Hopkins University Press
All rights reserved. Published 2013
Printed in the United States of America on acid-free paper
2 4 6 8 9 7 5 3 1

The Johns Hopkins University Press
2715 North Charles Street
Baltimore, Maryland 21218-4363
www.press.jhu.edu

Library of Congress Cataloging-in-Publication Data
Denny, Mark, 1953–
Lights on! : the science of power generation / by Mark Denny.
pages cm
Includes bibliographical references and index.
ISBN-13: 978-1-4214-0995-5 (hardcover : alk. paper)
ISBN-10: 1-4214-0995-X (hardcover : alk. paper)
ISBN-13: 978-1-4214-0996-2 (pbk. : alk. paper)
ISBN-10: 1-4214-0996-8 (pbk. : alk. paper)
ISBN-13: 978-1-4214-0997-9 (electronic)
ISBN-10: 1-4214-0997-6 (electronic)
1. Energy conversion. 2. Energy storage. 3. Renewable energy sources.
4. Power resources. I. Title.
TK1041.D46 2013
621.31—dc23
2012045324

A catalog record for this book is available from the British Library.

*Special discounts are available for bulk purchases of this book. For more information,
please contact Special Sales at 410-516-6936 or specialsales@press.jhu.edu.*

The Johns Hopkins University Press uses environmentally friendly book
materials, including recycled text paper that is composed of at least
30 percent post-consumer waste, whenever possible.

CONTENTS

ACKNOWLEDGMENTS

Thanks to Vincent Burke and Jennifer Malat at the Johns Hopkins University Press for ensuring the smooth transition of this book from conception to production. Thanks to Carolyn Moser, once again, for her expert copyediting. I am grateful to Dr. Charlotte Geier for permission to reproduce figure 2.4.

Two oil experts generously contributed their time and expertise to answering my questions on that important subject; I am much obliged to Dr. Alan McFadzean and Dr. Peter Ward for their contributions. I thank Dr. Carolyn Mattick for providing me with a copy of her paper on the history of energy production.

INTRODUCTION

This book is about energy and power—the kind that we need to heat our houses and light our streets, to get us from *A* to *B*, and to drive our industries. Power generation is a relatively recent problem historically because the human need for power was minimal until the Industrial Revolution. Nowadays it is, and is rightly seen to be, a formidable and perhaps overwhelming problem that we need to address, for the benefit of future generations as well as ourselves. In this book, I look into the different ways in which it is possible to generate power—to tap into one or another source of energy that is locked up somewhere, releasing it in a controlled and usable manner.

My approach is that of a scientist and engineer, not a politician or businessman. I am not inclined to excited hyperbole, though it seems to me that some people on the committed environmentalist wing of the current debate about energy sources do lean that way. Some folks at the other end of the spectrum are equally irrational, inclined toward equal exaggeration though usually less shrill. The result has been a heated debate (almost a pun—sorry) that spreads confusion and perhaps contributes to a regrettable apathy among the general public.

My aim in writing this book is to provide a readable exposition of the science and engineering of power generation, without raising your blood pressure too much. That is to say, I would be happy if you become engaged with the subject and (dare I say) energized as you digest the meat of this book, but I will not be expounding extreme or one-sided political views, or telling you what to do or think. I will be providing brain fodder for you to ruminate upon by presenting you with the story of our search for energy sources, the science behind each of the power generation technologies, and the facts of historical development. No politics—or, more realistically, no political agenda. (It is difficult to make a statement about our energy future without being political.)

Before saying more about our subject, let me provide you with a flavor of the approach to it that is adopted throughout this book. Here is a scientist-cum-engineer's quick glance at a much-discussed, promising and benign source of power, and its potential for solving the needs of humankind.

Humanity currently consumes a total annual average of 14–20 terawatts (TW) of power.[1] A terawatt is 1,000 gigawatts (GW); a gigawatt is 1,000 megawatts (MW); a megawatt is 1,000 kilowatts (kW). Switching on an electric kettle consumes a couple of kilowatts, and your electricity bill is likely expressed in kilowatts, or in kilowatt-hours (kWh), the energy equivalent of this ubiquitous power unit.

Here is a quick, back-of-the-envelope calculation to put into perspective a few conceptions that you might entertain concerning power generation. Most of our power and heat, and all of our light, come from a single giant thermonuclear power plant (a fusion reactor) in the sky. Our sun bathes the earth with electromagnetic waves, mostly in the form of visible light or invisible infrared radiation (heat). The amount of solar power that reaches our upper atmosphere is $1.35 \text{ kW}/\text{m}^2$; multiplying up by the cross-sectional area of the earth gives us about 172,000 TW. That is, the total amount of solar radiation that bathes the upper atmosphere of our planet exceeds our total power consumption by a factor of 8,600 (assuming the upper figure of 20 TW for current world consumption). Given this elementary fact, isn't it obvious that all our power source problems will go away if only we build up a massive infrastructure of solar power plants?

Not at all—although, as we will see in chapter 8, solar power plays an increasing role in our budget. First, only about 10% of the solar power that impinges upon the upper atmosphere actually makes it down to ground level. The rest is absorbed by the atmosphere or reflected off clouds. Also, half the time the power source is effectively switched off—we call such times "night" —and for only a few hours a day is the sun near its zenith. Many countries are too far north for efficient solar power generation; sunlight slants in at an angle and is rarely directly overhead—never so, for regions outside the tropics (north of the Tropic of Cancer or south of the Tropic of Capricorn).

1. The data are inconsistent; it must be extremely difficult to estimate the total of world power consumption. My data (and that which contributed to the graph) is drawn mainly from BP (2010) and IEA (2008).

Thus, the usable amount of solar power that reaches the surface is, in round numbers, "only" about 860 times the total power needs of our species.[2]

Let's keep going with this rough-and-ready look at solar power potential. The efficiency of a solar power plant is less than 1%. (Only a small percentage of the land area of a power plant is covered by solar panels, and the solar panels convert only a small fraction of the sunlight they receive into electricity. Interestingly, 1% is also about the efficiency of a photosynthesizing biological plant, converting solar energy into chemical energy.) This figure was gleaned from two of the world's largest solar power plants, both of which are in regions of very high *insolation*, or ambient sunlight levels. Let us agree on an average efficiency, for solar power plants all over the world, of 1%. Thus, our factor of 860 is reduced to 8.6. This figure shows (doesn't it?) that we don't need to bother with other, dirtier sources of power such as biofuels or nuclear plants: covering the world with solar power plants will provide us with more than eight times our annual power requirement.[3]

Not so fast. We cannot give over the whole surface of our planet to solar power plants. Most of the surface is ocean, and we need most of the land for other things, like cities and roads and agriculture and forests. We cannot realistically expect more than, say, 1% of our planet to be festooned with solar power plants, no matter how desperate we become for electrical power. Consequently, we obtain a value of about 8%—give or take—for the maximum possible contribution to our power needs from the sun. Further, the 20 TW figure for the power needs of our species is likely to rise significantly in the future, as the population rises and as Third World nations industrialize (see the chart). More people means more pressure for land, which is likely to depress my optimistic guess that 1% of the earth's surface (i.e., 3% of the land surface—about 1.7 million square miles) could be taken up by solar power plants.[4] However we cut and dice the numbers, if we are thinking realistically and not indulging in fantasy, then there is only one conclu-

2. Please understand that this is a ballpark calculation and so the numbers are very approximate. It is not possible, or even sensible, to attempt greater accuracy in estimating total human power consumption or useable solar power.

3. The figure for solar power plant efficiency comes from the Solar Energy Generating Systems (SEGS) facility in the Mojave Desert of California and the Alvarado I solar power plant in Extremadura, Spain.

4. Countries with large deserts, such as Australia, Saudi Arabia, and the United States, may be able to cover more than 3% of their land area with solar panels because desert land

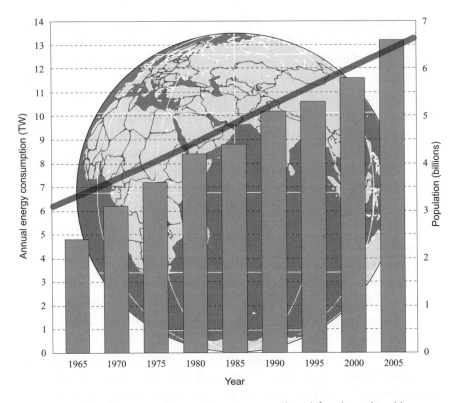

FIG. I.1. Annual world power consumption in terawatts (*bars, left scale*) and world population in billions (*line, right scale*), over a 40-year period from 1965. Note that energy
production increases faster than population, as Third World countries industrialize and
First World countries splurge. *Data from BP (2010).*

sion: solar power cannot form more than a small fraction of the total power
requirements of humanity.[5]

In the above reckoning I used only elementary math, please note, and a
broad-brush approximate approach that simplifies a complex subject and

---

is arguably not economically useful for farming anything other than sunlight. Also, proponents of solar power would argue that, in the foreseeable future, power plant efficiency will
increase. I unpack solar power research and development more completely in chapter 8.

5. I am guilty of mixing English and metric units here. At the risk of confusion, I will
reflect this regrettable power industry practice and continue to mix units throughout the
book. Generally, English units will be in the main text (except that I will refer to power in
kilowatts and energy in kilowatt-hours, because these are the units most people are familiar with from their utility bills) and metric in the sidebars and technical appendix.

yet is scientifically valid and incontestable. For example, the physical area of the earth provides an upper limit to the solar energy that it receives. This type of estimation (I hesitate to call it a calculation) was the hallmark of that well-known Italian American physicist and Nobel laureate Enrico Fermi, who once famously used it to estimate the number of piano tuners in New York City. The result is only a very approximate estimate, but to put it loosely, it is approximately right and not exactly wrong. It would be all too easy in this book to blind you with science; many of the matters we will look at are, in detail, very technical and complicated. Fermi's approach seems to be the wisest here—and apposite, given his significant contribution to one of our most important sources of power, as we will see.

In the first chapter I set the table with some clarifying observations on what we mean by energy and power, as well as set out the limitations imposed by physics and technology on how we convert between different forms of these. I will summarize the different ways in which we currently generate power and show how it is distributed. The rest of the book takes each of the power generation technologies (one or two per chapter) and explains how they work, with the aid of diagrams. Each technology will be subjected to Fermi-like analyses to test the claims made of them—for example, to test their potential for satisfying our growing need for affordable power, or to examine their environmental impact. To begin, I will summarize the history of energy and power generation to place the subject into context. For example, we will learn about the considerable efforts made by our predecessors to obtain coal and oil and to understand electricity and nuclear forces.

The existing literature on the subject of energy and power generation (its cost, sustainability, and environmental consequences) is truly enormous. Some of these books are excellent reads, while others would be better used as biofuel. Many are highly technical; many others have a political agenda or an environmental or business bias. Apart from academic texts, few of the existing books explain the science that underlies energy and power generation. Very few books explain this broad and complex subject in a manner that is accessible to the interested lay person. The book that you hold in your hands will fill this gap.

LIGHTS ON!

NEWTON'S LEGACY

"Energy is not a material thing," a physicist of the late nineteenth century might have said. Einstein would demur at this statement, and because of him, the physicists of later times think differently. Material or not, energy is a property of objects—a characteristic with a well-defined meaning. Our nineteenth-century physicist was not wholly ignorant about the subject, however; he knew that energy was conserved, that it could take on different forms, and that, like a genie or a spirit, it could change from one form to another. Sir Isaac Newton, who thought and calculated and wrote and sat under apple trees in the late seventeenth and early eighteenth centuries, knew nothing of energy. The exact, mathematical science that he bequeathed to us was expressed in the language of forces, not energy. After his death, Newton's laws of motion and of gravity would be elegantly reinterpreted in terms of energy by succeeding generations of gifted men.[1]

Yet the title of this chapter—which is about energy in its various forms, the transformation between them, and the distribution and storage of it—suggests a significant contribution by this great though difficult and unpleasant man. Why? Because modern engineers are taught the physics that was given to us by Newton, as well as that which emerged in later centuries. Newton's way of looking at the world is easier to convey to nonspecialists because it connects with things we have an intuitive feeling for, like force and momentum. Consequently, his methods are still used, particularly in the world of practical engineering, where Newtonian concepts are readily applied. For these reasons, in this book I will use Newtonian ideas—plus energy—to explain the physics of our subject. Please bear in mind, however, that the knowledge of modern-day engineers who design and build power plants, energy storage devices, and distribution networks comes from Joule, Carnot, Poncelet, von Helmholtz, Rankine, Einstein, and a thousand other physicists and engineers, as well as from Sir Isaac.

In this chapter I lay out our modern understanding of energy and power. This groundwork is necessary for a meaningful discussion of power generation, by various technologies and in the language of Newton, in later chapters. I avoid technical analyses but paint an accurate picture of the concepts and processes that we need later on.

Energy

One catchall definition of energy is "the ability to do work." Water flowing in a river can do work by turning a millstone; burning gasoline can do work moving an automobile. In the first case we have energy of motion—flowing water—and in the second case chemical energy, stored inside the molecules of gas. Through their ingenuity, engineers produce mechanical devices to convert these different forms of energy into useful work:[2] a waterwheel converts the energy of motion of a flowing river into the rotational energy of a millstone; an internal combustion engine and a crankshaft convert the chemical energy of gasoline into the energy of motion of an automobile.

Energy is related to force (which is why we can use the language of Newton to describe the subject). The work done in moving a mass is just the energy expended on the task: it is the force applied to the mass multiplied by the distance it moves. For example, the energy you expend in dragging a heavy trunk across an attic floor is the force you apply to the trunk multiplied by the distance you drag it across the floor. Increase the force, or increase the distance, and you increase the energy that you expend, and so the work that you do.

There are many types of energy, and the connection between them is well understood by physicists, who can write down formulas that tell us, for example, how to convert mechanical energy into heat. A moving mass possesses mechanical energy; the mass may be moving along a straight line, or it may be stationary and rotating about an axle. Both freight trains and flywheels posses such *kinetic energy*, or mechanical energy of motion. *Potential energy* is the energy possessed by a mass by virtue of its position. Thus, a compressed spring and a rock raised up from the ground both possess potential energy. We can see that this is so by considering what happens when the spring and the rock are released. The spring will jump in the air or cause a mechanical toy to move; the rock will fall to the ground (thus gaining kinetic energy) with a thud (acoustic energy). Here we have another example of energy being converted from one type to another. Again, physicists can

A LIGHT WORKOUT

To shed some light on the common units of energy, let's consider a light bulb. This little exercise will give us a feeling for the magnitude of energy in its different forms. The energy consumed by a 100-watt (W) light bulb that is switched on for one hour is 0.1 kilowatt-hour (kWh). Here are some equivalents: The same amount of energy would enable an adult man to climb about 300 feet up a ladder (assuming he weighs 180 pounds and converts only 20% of the energy into useful mechanical work). The same energy expenditure would bring a stationary half-ton truck up to a speed of 13 mph (assuming 15% fuel-to-wheel efficiency). The same energy could bring a little less than 5 ounces of water up from room temperature and cause it to boil. The same energy could power stereo speakers to play music loudly in a moderate-sized room for 8 hours.

Our estimation of energy consumption is somewhat biased by our sensory perceptions. The quietest sound we can hear contains much less energy than the dimmest light we can see. The energy needed to raise water to 100°C is much less than the energy needed to cause it to boil at 100°C.

write down equations that tell us exactly how to convert potential energy into kinetic energy. For example, they know how fast a rock will hit the ground if it is released from a specified height above the surface. Kinetic and potential energy, and the conversion from one to the other, is illustrated in figure 1.1.

Thus, energy falls into one of two categories: the kinetic energy of motion and the potential energy of position. Chemical energy is a kind of potential energy.[3] It takes a lot of energy to create the molecules of, say, gasoline, and this stored energy is released when the gasoline is burned. Gunpowder, nitroglycerine, firewood, and a myriad of other materials possess internal chemical energy that can be released in a useful manner: they can do work. Heat is a form of kinetic energy,[4] sometimes called the lowest form. Heat is like money—a common currency in which we trade different commodities (in this case, forms of energy) and compare their values. All other forms of energy can turn directly into heat. Thus, a meteorite falling to earth converts kinetic energy and gravitational potential energy into heat: it arrives at the surface hot and has heated up the air it passed through. The chemical energy stored in gasoline is converted into thermal energy, which in turn is converted into the mechanical energy of moving pistons. The laws of thermo-

FIG. 1.1. A space shuttle launch: chemical potential energy (liquid hydrogen and liquid oxygen fuel) is converted into the kinetic energy of the speeding rocket, in order to overcome the gravitational potential energy of the Earth. *Photo courtesy of NASA.*

dynamics and mechanics, and the detailed engine design characteristics, tell us how much horsepower we can extract from a liter of gas. In a nuclear power plant, nuclear energy is converted into heat, which is then in turn converted by turbines into electrical energy.[5] Two simple examples of energy conversion are shown in figure 1.2.

In our power generating plants, of whatever type, energy is converted from one form to another several times before emerging as electrical energy. Consider, for example, a dam. Historically built to contain water,[6] most dams today serve another purpose—providing hydroelectric power. First,

Pendulum-and-spring apparatus

(a)

(b)

Air gun

(c)

(d)

(e)

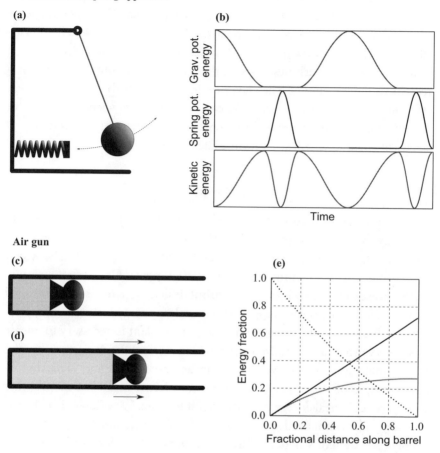

FIG. 1.2. Two examples of energy conversion—a pendulum and spring and an air gun. (a) This pendulum-and-spring apparatus stores potential energy in a spring and gravitational potential energy in the pendulum bob. The bob also exhibits kinetic energy of motion. (b) Analysis shows how the energy of this apparatus is divided up and how the division changes with time. In this hypothetical case, with no energy dissipated by friction, the sum of all three forms of energy is a constant. (c) Another example of energy conversion, an air gun, showing a pellet at rest in the barrel before the trigger is pulled. Behind the pellet is compressed air, which has a lot of potential energy. (d) After the trigger is pulled, the compressed air is released, and the pellet accelerates along the barrel; potential energy is converted into kinetic energy. (e) A graph of energy fraction vs. pellet position along the barrel: *dotted line*, the potential energy of compressed air; *black line*, energy lost to friction, mostly as heat ; and *gray line*, the pellet's kinetic energy. This air gun is about 30% efficient because 30% of the initial potential energy is transferred to the pellet.

water levels build up in front of the dam, and the water acquires gravitational potential energy. This potential energy is converted into kinetic energy when the water flows down into the hydroelectric turbines, which are usually positioned inside the dam near its base. The flowing water gives up most of its energy to the turbines, causing the large turbine rotors to spin fast—rotational kinetic energy. The clever design of such turbines leads to high efficiency, but no energy conversion can be perfectly efficient, and some of the water energy is wasted as heat. Finally rotor energy is converted into electrical energy; we will see how in a later chapter.

Power

Power is the flow of energy—the rate of change of energy. If you expend 100 J of energy dragging that trunk across that attic floor, for every second that you drag it, then you are expending energy at the rate of 100 J/s, which means that you are exerting 100 W (0.1 kW) of power.[7]

In some cases, energy is easy to store but difficult to release in a controlled manner, as a steady power supply. Thus, nuclear energy is stored compactly as mass, but a carefully designed nuclear power plant is necessary to release this energy in a steady flow. In other cases, energy is difficult to store, but the flow of that energy—its power—is readily achieved. Electricity is *the* important example of this type. It is the most efficient means of distributing energy, yet electrical energy is very difficult to store. We will see that electrical energy that is not needed immediately is converted into different types of energy (potential or kinetic) before being stored. Other forms of energy can be stored easily and also can be made to flow easily, providing steady power. Water, by virtue of its weight and movement, has historically been our most important source of power for this very reason.

Two examples of natural energy flow—of power—are shown in figure 1.3. These examples illustrate the potential for, and the problems of, human power generation. Niagara Falls (fig. 1.3a) expends power at an average rate of about 1.3 GW. We can readily estimate this value from knowledge of the flow rate and the drop in height of the water. Because of seasonal variations in the flow rate as well as human intervention in allowing the river to flow freely, the power of the falls varies. Falling water is a key component of human power generation because it is readily exploited. Indeed, the Niagara River is diverted from the falls, on both the American and Canadian sides of

POWER UNITS

Consider a half-ton truck with a 230 horsepower (hp) engine. Such an engine can power 1,715 incandescent light bulbs (each of 100 W). A horizontal stream or a water channel with a cross-sectional area of 5 m² (say it is 1 m deep and 5 m wide) flowing at 12 mph may generate the same amount of power, assuming an efficiency of 50% for the waterwheel or turbine that converts the water power into mechanical or electrical power. If the water is falling through the same water channel, as in a waterfall or inside a hydroelectric dam, instead of flowing through it horizontally, then it can generate the same power by falling only 0.9 m—about 3 feet. (I am mixing up English and metric units here—horsepower and watts, feet and meters. This practice reflects the real world. As a scientist I would prefer to consistently use metric, but the engineer in me knows that people like some of the older units.)

---

the border, for the purposes of hydroelectric power generation. Up to 75% of the flow is removed above the falls and returned to the river below the falls.[8]

The second example of natural energy flow is that of lightning (fig. 1.3b). A typical lightning bolt transfers about 140 kWh of energy from a cloud to the ground in a fraction of a millisecond and so represents a prodigious—albeit brief—rate of power: typically 4 TW. We do not use lightning as a source of power for a number of obvious reasons, but we do choose to transfer our energy from one place to another via electricity. The distance covered by a lightning bolt may be only a few hundred meters; the controlled flow of electricity along uninsulated conductors can be over hundreds of kilometers.[9]

Energy Conversion

In engineering, *transducer* is the generic name given to a device that converts energy from one form to another. No transducer is perfectly efficient; there is always wastage to a greater or lesser degree. In some practical cases the efficiency of transducers can approach 100% as technology improves—for example, with design improvements, friction reduction, or size reduction (to reduce the energy absorbed by moving parts). In many cases, however, it is impossible even theoretically to convert energy perfectly. An important

FIG. 1.3. Two examples of natural power. (a) Horseshoe Falls, the largest of the three Niagara waterfalls. This image illustrates the power of falling water. Except during peak tourist visiting hours, much of the Niagara water is diverted to hydro power stations. (b) Electrical potential energy that builds up in clouds discharges to the ground. A typical lightning bolt represents about a quarter of the power used by humanity—but only for a tenth of a millisecond. *Photos by (a) Ujjwal Kumar and (b) NikoSilver.*

class of transducer that fits into this latter category is the *heat engine*. Heat engines can be considered abstractly (physicists will recognize the *Carnot engine* here, an early thermodynamic theoretical tool that demonstrates the second law of thermodynamics). However, many real engines fall under the same umbrella: steam, diesel, and gasoline engines are all heat engines, and so they are limited in efficiency by the laws of physics, not just by practical engineering considerations.

There are many familiar examples of transducers. The light bulb is a common transducer that converts electrical energy into light, a form of electromagnetic (EM) radiation. The reverse process is achieved with a photovoltaic solar panel. We have seen already how inefficient solar panels are; light bulbs are almost as bad. The old incandescent bulbs convert about 5% of the input electrical power into EM power; fluorescent lights are better, at about 20%. Metal halide lamps are better again, around 25%, while LEDs can be up to 35% efficient and low-pressure sodium lamps 40%. An electric heater converts over 95% of the input electrical energy into heat, another form of EM energy. The heater is so efficient because heat is the desired output, whereas it is an unwanted by-product of most conversion processes. A home gas furnace is another example of a transducer, this time turning stored chemical energy into heat. Again, heat is the desired output, and so heat production is not a source of inefficiency—quite the opposite. Such furnaces are about 95% efficient. Oil furnaces are not quite so good, at 65%. The steam turbine is an example of the reverse transducer, turning heat into electricity; turbine efficiency is usually between 45% and 60%.

A crankshaft converts one type of mechanical kinetic energy into another: it turns reciprocating linear motion into rotary motion, and vice versa. It is not a heat engine; thus, its efficiency is not theoretically limited. Gear trains are another example of mechanical transducers that can convert one type of mechanical energy into another. Because such devices are not heat engines and are not limited theoretically in their efficiency, good design and good manufacture can result in very efficient conversion.

An automobile engine is a transducer for converting chemical energy into rotational mechanical energy, which it does at about 25% efficiency. (The lower efficiency of the whole automobile, 15%, is due mostly to drive train inefficiency.) Electricity generators convert chemical energy into electricity; we will look at the different types of generators in chapter 3. Dynamos convert rotational mechanical energy into electricity, while electric motors do the reverse. The little hub dynamos that power bicycle lights vary in their efficiency, depending upon speed, but peak at about 65%; automotive alternators get up to 90% efficiency—that is, they turn up to 90% of their input mechanical energy into electricity, which is used to charge up the automobile battery. Electric motors vary in efficiency between about 40% and 90%; larger motors are more efficient than smaller ones.[10]

Figure 1.4 shows a transducer loop—which is impractical, of course, because of the waste from the inefficiencies of our technology and of heat engines.

There are many other types of transducers. A microphone converts acoustic energy into electricity; a speaker does the reverse. Windmills and water-wheels convert the kinetic energy of wind and running water (traditionally) into the rotational kinetic energy of a moving millstone. Antennas are transducers; they are like electric lights or heaters, in that they convert electrical energy into EM radiation, except that antennas can also work in reverse, converting microwave or radio signals into electrical signals. Muscles convert chemical energy into mechanical energy, as do internal combustion engines.

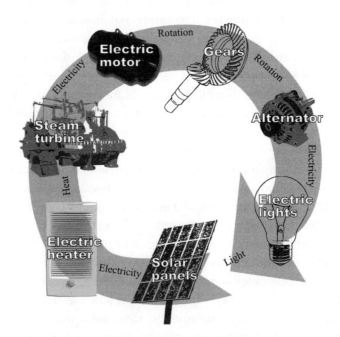

FIG. 1.4. Energy transformation. If all these transducers were perfectly efficient, energy would circulate around this loop endlessly, without loss. In practice, because of the efficiency values of each component, 99.8% of the energy would be lost after just one cycle.

Hydroelectric dams convert the gravitational potential energy of water into electricity, and batteries convert chemical energy into electricity.

Note that many of my transducer examples involve electricity. This reflects our real-world dependence on this form of energy and, particularly, power. It turns out that

- electrical energy is difficult to store, and
- electrical power can be transported from *A* to *B* readily.

I alluded to these phenomena earlier; they are important and are discussed in depth in chapter 3. For now, we need only know that they apply in the real world and that we must live with the consequence, which is this: no matter how we generate energy on a large scale (at the level of big utility companies and of nations), we must find efficient ways to convert it into electricity because that is the most economical way to distribute energy over long distances.

Power Distribution

The large-scale distribution of power—across municipal boundaries, across nations, across continents—varies in efficiency depending upon its form. Such large scales make the distribution of most sources of power expensive or impractical. For example, flowing water can be moved tens of kilometers via aqueducts—but no farther than this, and only if the topography is right. The Romans were very skillful at distributing water in this way, though in their case the water was usually for drinking or irrigation, not for generating power. (We will encounter an impressive exception in chapter 2.) Moving water over longer distances for power generation would prove to be wasteful (water flows only downhill, thus depleting its potential energy) and expensive in terms of infrastructure.

Oil contains chemical energy, and today there are enormous pipelines and supertankers that transport this fuel across the globe. (We can regard such transportation as power distribution as well as energy distribution; the energy is flowing.) But oil, coal, firewood, food, and all the other sources of stored chemical energy that we need in order to live our lives and power our industries are expensive to move because they are heavy, like water. Because they are heavy (and, in some cases, toxic), we do not transport such chemical energy sources over long distances unless we have to, as with oil and coal.

What about heat? Heat weighs little,[11] but it is difficult to transport more than a few tens of meters (as in a power plant) because it leaks. I am unaware of any heat pipelines or "calorie cables" snaking their way across the countryside or throughout cities.

We are obliged to transport oil and coal over large distances because it is more economical to do so than to burn the oil or coal at the source and then transport the power it generates as electricity. (It is not always possible to place power plants at the oil wells or coal fields: for example, many oil fields are under the sea.) And we move oil and coal because of limitations in our technology: our automobiles are powered by oil derivatives, and our power plants rely upon oil or coal as fuel.

In general, though, the weight of most sources of chemical energy makes them too expensive to transport. Even if we could reduce the transportation costs, there are other reasons why electricity is the favored means of distributing power. First and foremost, it is fast. Electrical power is carried via cable at about the speed of light.[12] Second, it is compact and can be squeezed into thin cables. These cables can be distributed widely, then split and redistributed, throughout a city and throughout a house. These properties make electricity the favored form of power distribution; it is far more convenient than any other form.

Speed of distribution matters to the companies that generate power for a city or a nation because of the economics and technology of power generation. We will see that most of our power plants work best—most efficiently—when they are producing power at a constant rate. Yet the demand for power is far from constant. We heat our homes more in winter than summer. We work during the day and sleep at night, so power needs are reduced at night. We get up in the morning and switch on the toaster or coffee maker, causing demand to surge. Power demand thus varies over many timescales: annually, daily, and from minute to minute. There are spikes in local power consumption for any number of reasons: during the Super Bowl in America, at halftime during televised soccer games in Europe and South America, after a popular soap opera when millions of teakettles get switched on in Britain.

All these myriad sources of transient demand for electrical power present themselves as a random fluctuation in the output that is required from a nation's power grid. A single generating plant might be required to supply electrical power at a rate that fluctuates slowly with the seasons, fluctuates more quickly with the hours of daylight, and also fluctuates randomly from one minute to the next. Figure 1.5a provides an illustration of the demand

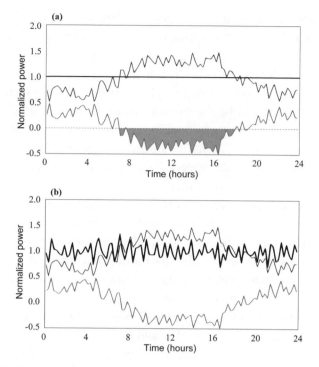

FIG. 1.5. An idealized power management system. (a) A power station supplies a city; the demand (*upper jagged line*) varies not only daily but also randomly from minute to minute. This demand is met by the basic power supply only up to a certain level (here shown as 1.0). To this basic supply is added a varying supplementary supply (*lower jagged line*) by storing and releasing energy as required, so that the total power supply (*bold horizontal line*) is constant. The shaded area shows when stored energy is being expended to meet high demand; at other times stored energy is being increased. For this simulation, the day has been divided into 15-minute intervals, and the response to changing demand is idealized as instantaneous. (b) A slightly more realistic case in which there is a 15-minute response time between demand and supply. Note that there is now a fluctuation in total supply (*bold line*), which a utility company would want to minimize, for reasons discussed in the text.

from a power plant. This fluctuating demand causes problems because the grid of power plants supplying a region or nation works best when generating constant power. If a generating plant produces power to meet *peak* demand, then much power is wasted. If, on the other hand, a generating plant produces enough power to meet *average* demand, the waste is reduced, but there will be times when demand is not met unless power is held back during low-demand periods and then supplied during high-demand periods. This option is the one usually adopted, because it is more economical and more practical than attempting to meet peak demand for power.

In figure 1.5a we see also the manner in which a power plant must respond to local demand so that it is able to maintain a constant supply. The plant manages its power by using surplus to store energy. (Exactly how this storage is achieved will be the subject matter of the next section.) When needed, this extra stored energy is then converted into electrical power and distributed as required. The result, idealized in figure 1.5a, is a generating plant that produces a steady supply of electrical power—just enough to meet the average demand of its customers.

More realistically, a power plant cannot respond instantaneously to changes in demand although, as we will soon see, it can respond quite quickly. Assuming a 15-minute response time, the required supply for our hypothetical generating plant is shown in figure 1.5b. Now the demand for power is not reduced to a perfectly constant level; the managed power supply fluctuates randomly but much less than it would without power supply management via energy storage. In order to ensure that it is able to meet its customers' power demand requirements, the generating plant would have to increase its level of power production (assumed to be constant) to match the peak of these reduced fluctuations. Clearly, it is in the economic interest of the power supplier to minimize the level of these residual fluctuations.

This minimization can be achieved by pooling the resources of many generating plants. Let us extend our hypothetical example to 10 generating plants supplying 10 cities. We will assume that the average demand of each city is the same and the daily variations in demand are the same; only the minute-to-minute fluctuations change from one city to the next. Each generating plant supplies the same average amount of power. By pooling resources, the generators limit the statistical fluctuation in demand, to about 30% of the fluctuations shown in figure 1.5b.

This simplified simulation of the power demand-and-supply situation for a large power producer demonstrates that power management requires some considerable sophistication. It is wasteful to simply supply the peak power demand, and yet producing a lesser amount requires a careful and quick response to minute-by-minute changes in the amount of stored energy, and its release back into the power grid, in order to meet the ever-fluctuating demand of customers.

The economics of power generation has led to a considerable maturing of generation and distribution strategy and technology, as you might imagine. In fact, it is an ongoing process; the latest ideas and capabilities go under the name of *smart grid*, and they represent the contribution of digital technology to the current power management schemes. The idea is to further improve efficiency and increase the resilience of an electricity grid to fast-changing circumstances. For example, as customers develop their own solar power supplies, they will be able to sell excess power back to the grid, with bi-directional metering managing the process automatically. Variability of power—likely to increase in the future, as solar and wind power become more significant—will be smoothed out faster, by local energy storage and also perhaps by *local control switching*.

Local control switching involves the power supplier's checking domestic usage—for example, by monitoring and restricting the times at which high-power household equipment (washing machines, air conditioners, water heaters) is used. Half the power used in average American households goes to consumer electronic goods, such as TVs and computers; local control switching would monitor usage and automatically hibernate equipment that is not in use. These ideas are implementable; the civil liberties issues remain to be sorted out. The idea is to reduce consumption during peak times and move it off-peak, thus reducing the likelihood of widespread brownouts.

An example of smart-grid thinking is the vehicle-to-grid (V2G) idea, whereby electric and hybrid plug-in cars charge up off-peak and sell excess power back to the supplier at peak times. V2G will help smooth out the demand placed on a utility—and it is easier than it sounds because most vehicles, studies show, spend 95% of their time sitting in the garage.[13]

The smart grid will be capable of responding quickly to a sudden loss of supply—for example, due to a fire or mechanical breakdown—or to disruptions caused by lightning or sabotage. It will improve cost-effectiveness by switching suppliers if there is a change in price of one or another source of power (e.g., a drop in the price of solar power due to a hot summer, or a rise in oil prices due to pipeline leakage or political machinations). It can mitigate the daily fluctuations in supply across continents by shuffling power from one time zone to another. To some extent, the grid is already managed along these lines (pardon the pun), but smart-grid technology will enhance and speed up the response of suppliers to increasingly variable demand.

One final comment on power distribution is called for. So far, I have discussed power distribution on a large scale for the purpose of demand

POWER CURVES

I have hinted at the complexity of matching power demand with supply. This is a real problem that utility companies have had to address because it affects their bottom line in a big way. Suppliers must allow for fluctuations in demand but do not want to provide more power than is necessary. To help them assess demand they often plot data that they have gathered on power usage as a *load duration curve*. The vertical axis of such a curve represents power demand or load (which, as we have seen, varies over time). The horizontal axis represents the fraction of time that a given load is required or exceeded (the *capacity utilization*).

Consider as an example the imaginary village of Sparksville, which receives electricity from its own municipal plant. The good citizens consume electrical power over a four-day period as shown in part (a) of the figure. The average load for Sparksville appears to be constant during this period but fluctuates about this value from minute to minute. The corresponding load duration curve is shown in part (b) of the figure. From this curve we see that for this four-day interval, half of the time 3 MW of power was being drawn from the municipal plant, but for 10% of the time demand exceeded 4½ MW. (Over the four days, Sparksville drew power at an average rate of 3 MW, consuming 288 MWh of energy.) Such information enables the Sparksville power plant managers to decide how much power they should produce. Combining the load duration curve with a *price duration curve* (which plots the price of electricity instead of load) allows them to see how best to maximize profits for a given supply of energy.

management—to even out the required supply in the face of variable demand. At a smaller scale—say, for a town or a factory—power distribution is necessary for a somewhat different purpose: to bridge between a grid outage and the supply of backup generator power. (We will see an interesting example of such bridging in the next section, with the world's biggest battery.) On the smallest scale—say, for the power supplied to your desktop computer—power sometimes needs redistributing to ensure its even flow. The evenness of power flow is measured by *power quality*, and it is an important factor for many types of electrical equipment, which do not like to receive spikes of excess electrical power or loss of power, however brief. The most effective means of distributing power on these three scales (grid management, bridging, and power quality) depends upon the scale.

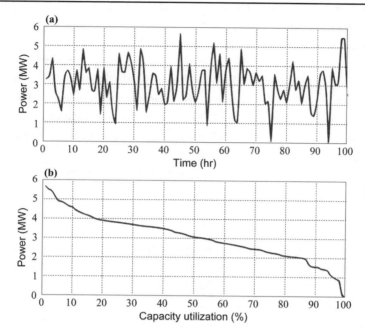

Electrical power consumption in Sparksville over approximately four days. (a) The average value fluctuates randomly, as in figure 1.5, though here for simplicity I assume no underlying daily variation. (b) The same data presented as a load duration curve. Capacity utilization is the fraction of time that a given power level is required or exceeded. Thus, 1 MW is required almost all the time, but 3 MW is needed only half the time (50% utilization).

Energy Storage

Why is electrical energy hard to store? We know that it can be done. Electric batteries power household equipment and an increasing number of cars. Major utility companies need energy storage to manage fluctuating demand, but of course we need energy storage at a more personal level as well. Some of the storage methods discussed here work well on a small scale, while others are better suited for large-scale applications. We will begin by looking at batteries. They are effective and are continually being made more so, but they are expensive. Another device for storing electricity is the capacitor, which stores energy by separating positive and negative charges (with an insulating material between the charges to prevent discharge). Capacitors can respond much more quickly than batteries and are useful in certain restricted circumstances, as when covering a sudden loss of power to a critical piece of equipment before a generator can switch on. But capacitors do not last long before discharging completely.

Consider, as a quirky example of a natural capacitor, a cumulonimbus thunderhead cloud. I am not suggesting that clouds could really be exploited as a means of storing electricity, but this example illustrates the severe limitation of a capacitor—it is a low-density form of stored energy. (The energy stored in clouds is huge, but only because clouds are huge. A capacitor the same size as a battery stores much less energy than the battery.) Capacitors—engineered ones, not my pie-in-the-sky example—are being actively developed for use in the automobile industry, but their characteristics are not suited to utility-scale storage.[14]

BATTERIES

The ubiquitous 1.5-volt AA battery is the familiar face of an industry with annual sales of $50 billion. There are two main families of batteries: primary (disposable) and secondary (rechargeable). Within each family there are a couple of dozen different types. Each type works with a particular electrochemical reaction that produces a small potential difference between the two terminals (1.5 V is typical). The characteristics of the different types vary, so that some are better suited to a given application than are others.

Disposable batteries are less expensive than rechargeables, as a rule. They can be stored for a long time, but they slowly self-discharge[15] and cannot be recharged: the chemical reactions are one-way. They have a higher energy density than do rechargeable batteries, but they do not perform well when supplying a large current. In fact, their efficiency varies with the electric current they supply; they are most efficient when the current drawn is small.

The least expensive and therefore one of the most common disposable batteries is the zinc-carbon type. The two electrodes (positive and negative)[16] are separated by an *electrolyte* that consists, in this case, of a paste made from zinc chloride ($ZnCl_2$) and ammonium chloride (NH_4Cl) dissolved in water. The electrolyte permits flow of electrons (i.e., of electric current) between the two terminals when they are connected; in other words, the electrolyte is a conductor when the circuit is closed. Carbon-zinc batteries have been around for a century, during which time they have been improved to the extent that their lifespan has improved fourfold, all else being equal. As is typical for batteries, the lifespan varies with use (frequency of use, load) and with ambient conditions (temperature, humidity). Zinc chloride batteries are a heavy-duty version of the zinc-carbon type. There are also alkaline batteries, and lithium and silver oxide batteries, among many others. They differ in both electrode and electrolyte material (and therefore differ in the electrochemical reactions that generate the potential difference between electrodes).

Zinc batteries find general use in low-current applications. Alkaline batteries are more expensive but have a higher energy density and a longer shelf life. (However, they do more damage when they leak, due to the corrosive potassium hydroxide electrolyte.) Lithium batteries are found in pacemakers and digital cameras. Silver oxide batteries are restricted to button cells, those little pill-shaped batteries that are fitted into wristwatches and some calculators.

When not in use, rechargeable batteries lose their charge more rapidly than do disposable batteries. They can be recharged between 500 and 1,000 times; the exact number of cycles depends upon how they have been used. Deep discharges shorten the lifetime. More so than common disposable batteries, rechargeable batteries often contain toxic chemicals and require careful disposal when the battery is dead. The common nickel-cadmium (NiCd or NiCad) battery finds application in toys and portable electronic devices; lithium-ion batteries are used to power laptops; lead-acid batteries start up conventional automobile engines; and nickel-metal halide batteries are used to power electric cars and hybrids.

Much effort has been put into the design of all batteries, perhaps none more so than the common lead-acid car battery, which has to operate in a harsh environment. (It is subjected to a lot of vibration and frequent temperature changes.) Technically, the small batteries that we use in our flashlights and remote controls are *cells*—a cell being a single anode-electrolyte-cathode unit. Car batteries however consist of six cells, each producing 2 V and connected in series, to boost the potential difference between the end electrodes to 12 V. The electrolyte in car batteries is a sulfuric acid solution, so again they give rise to disposal issues when finally they die.

The fundamental problem that prevents batteries from being used on a large scale—say, for utilities—is that only a small potential difference (1½–2 V) can be generated by a cell. To boost the juice to utility proportions requires linking together a large number of individual cells, but because the energy density of such cells is small, compared with most other sources of energy, the resulting high-voltage battery is very large indeed. Consider BESS, the Battery Energy Storage System that is used in Fairbanks, Alaska, to smooth out the frequent outages in that part of the world. It takes a few minutes for generators to come online, whereas electric batteries (which, by the way, are very efficient) are much quicker, so that BESS covers the power needs of Fairbanks for a few minutes until the generators kick in. This coverage can be critical in an Alaskan winter, which can see temperatures of 50 below. BESS can supply power at the rate of 40 MW, but only for 7 minutes (or 27 MW for 15 minutes). This output makes BESS the world's biggest battery, and it truly is huge. The 13,760 NiCd cells weigh a total of 1,300 tons and occupy an area of 21,500 ft^2. Given these figures, it is easy to see why batteries are generally small, with few large-scale applications (the most common large-scale applications are in telephone exchanges and computer data centers).[17]

PUMPED HYDRO

For many years now, the most popular method of large-scale energy storage has been pumped water storage. Indeed, it is the only economical storage method. Pumped water storage amounts to about 3% of instantaneous global power-generating capacity (90 GW, of which 20 GW is in the United States, spread over 38 hydropower plants). The technique is very simple, and quite efficient, but the engineering is big and expensive and requires special geology. The idea is to store energy during off-peak times by pumping water uphill to an elevated reservoir above the hydroelectric generators. During periods of peak demand, this stored potential energy is released and powers turbines to generate extra electricity. The efficiency is 70%–80%, meaning that about three-quarters of the energy expended in pumping the water uphill is later recovered as electric power. Losses are due to pump inefficiency and dissipative forces within the water (friction, viscosity), plus evaporation while the water is in storage. The whole process of pumping and recovering the electric power is thus a net consumer of power (because efficiency is less than perfect), and yet such energy storage generates revenue because of the difference between peak and off-peak power rates.

Special geological requirements limit the use of pumped water storage. Only a fraction of sites are suitable for hydro power generation; of these only a fraction are suitable for pumped water storage. A reservoir is needed, and it must be close to the hydro plant (to reduce pumping losses, which increase with distance pumped) yet at significantly higher elevation (to maximize stored potential energy). The reservoir sites adopted are usually at least 200 m higher than the hydro plant. A clear example of favorable pumped water geology is shown in figure 1.6.

As you might imagine, there are significant engineering challenges in

FIG. 1.6. The Seneca Pumped Storage Generating Station, in Warren, Pennsylvania. Constructed in 1970, the reservoir holds 280 million cubic feet of water; it is 60 feet deep and is 800 feet above the hydro plant (the Kinzua hydro dam at bottom left). Factor in the density of water and the acceleration due to gravity, and we find that the storage capacity is 5.2 GWh. This is enough energy to keep the power station running at peak capacity (443 MW) for about 12 hours. *Photo by U.S. Army Corps of Engineers.*

building pumped hydro sites. The physics, however, is easy. We can apply a quick Fermi-type back-of-the-envelope calculation to roughly estimate the energy storage capacity of a given site by multiplying the weight of stored water by the height above the hydro generators and by the pumping efficiency (as I did in the caption to fig. 1.6). A recently completed very-large-capacity pumped hydro plant is the Tianhuangping facility in China. Here, six Francis turbines pump water 880 m up to a top reservoir with capacity of 6.8 million m³. (When this mammoth project was completed in 2004, the construction bill was $900 million, which shows the scale of this pumped water facility.) Assuming 80% efficiency, the calculation tells us that the facility can store up to 13 GWh of useable energy (2% of China's daily consumption). This quick estimate is very close to the published figure.

The same hydro plant turbines that generate electricity can be operated in reverse to pump water up to the reservoir when needed.[18] The turbines can be brought up to operating speed—in either direction—quite quickly; in general, the reaction time of a pumped hydro plant is 1–4 minutes. That is, when required to stop generating electrical power and instead store potential energy, or vice versa, the plant can respond in a few minutes. Such a quick response time greatly helps with load-balancing. Because of the varying demand for electric power, it is normal for pumped hydro plants to follow a daily cycle of generating and storing.

COMPRESSED AIR

Pneumatic tools are used in certain niche areas—for example, for dentists' chairs and drills. Historically, compressed air has been used to power torpedoes, engines in mines (where emission issues ruled out steam or internal combustion engines), and office messaging, among other things.[19] Today, instead of being used in small-scale power applications, compressed air is considered mainly for large-scale energy storage. Indeed, for grid energy storage it is the only realistic competitor of pumped hydro. Compressed air energy storage (CAES) possesses a couple of features in common with pumped hydro: both build up potential energy (compressed air, elevated water), and both require special geology.

The idea behind compressed air storage is illustrated in figure 1.7. The air is stored underground, in a suitable structure such as an old salt mine, an aquifer, or a cavern. The key requirement is that the storage structure be airtight because the air inside it is highly compressed (in one case, to more than 100 atmospheres). Unfortunately, not many such sites have been identified, so that at the time of this writing there are only three large compressed air storage facilities in the world. Another challenge for CAES lies in the operating characteristic of turbines: they lose efficiency when the air pressure changes. This fact militates against a small storage reservoir; it has to be very large, so that pressure changes are small when compressed air is released. More fundamental is the thermodynamics of air compression and expansion. Unless the change in pressure and volume is very slow, air will cool as it expands (think of the ice that forms at the outlet of a gas cylinder). Air that drives a gas turbine needs to be warm, as we will see, and so it is necessary to heat the air that is released from the reservoir before using it to drive a turbine (as suggested in fig. 1.7). Heating the air reduces efficiency. In practice, efficiencies of CAES facilities seem to be a little above 50%, meaning that about half of the energy used to compress the air and later to reheat it is recovered as generated electricity. Industry brochures suggest that efficiencies up to 70%–80% may be attainable in the foreseeable future.

The world's first CAES facility was constructed in Huntorf, Germany, in

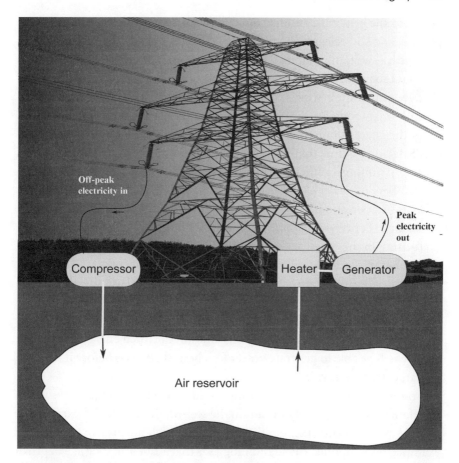

Off-peak
electricity in

Peak
electricity
out

Compressor

Heater

Generator

Air reservoir

FIG. 1.7. Compressed air energy storage. During off-peak periods, electric power is used to compress air, which is stored in an underground reservoir. During peak periods the compressed air powers a generator that supplies electricity back to the grid. The need for a heater limits CAES efficiency.

1978. It generates 290 MW of power when called upon to do so and can maintain that output for two hours, after which air pressure is reduced to the extent that the 10 million ft³ reservoir needs refilling. The peak operating pressure of the compressed air is 70 atmospheres. The world's second CAES facility, providing 110 MW, was built in McIntosh, Alabama, in 1991; and the third and largest facility, still under construction, is in Norton, Ohio; it will provide 2,700 MW of electric power. The first two facilities make use of old salt mines, while the third takes advantage of an abandoned limestone mine.

We can estimate the length of time that such CAES facilities can provide power to the grid with another quick calculation. This one involves a little physics, and so I relegate the technical details to the appendix. The point is that, with a sweeping simplification, we can come up with ballpark estimates without too much math. The sweeping simplification is that the air does not change temperature when it is compressed or when it expands. This assumption is wrong but not so wrong as to render the estimate worthless. From the analysis presented in the appendix, and the data for the Huntorf CAES facility, I find that the compressed air stores about 580 MWh of energy. Assuming an efficiency of 50% for the plant, this facility can contribute 290 MW of power to the grid for about an hour. Thus, my calculation in this case underestimates by a factor of two—not bad for such a simple calculation. Fermi would approve, I feel.

Turning to the massive CAES site at Norton, Ohio, a similar calculation shows that the compressed air stores up to 420 GWh of energy. (Here I have made use of the following data, gleaned from the Norton website: the storage volume of the reservoir is 350 million ft³, and the reservoir can withstand pressure of up to 1,600 psi.) Given the design output power of 2,700 MW and the same 50% efficiency as before, this amount of stored energy means that the site will be able to generate power for about three days before it needs to recharge its air reservoir.

An interesting application of compressed air energy storage, currently in its infancy, involves storing the air on the seabed, in inexpensive expandable bags. Water pressure (i.e., depth) determines peak air pressure. Thus, for example, my simplified calculation suggests that a 1,000-m³ bag at a depth of 400 m could store about a megawatt-hour of energy.

FLYWHEELS

Rotational kinetic energy has an advantage over the other form of kinetic energy—linear motion—in that it takes up relatively little space. High rotational speed yields a high energy density, and so modern high-speed flywheels are beginning to supersede batteries in certain applications as the preferred method of storing energy.

Traditionally, flywheels have been used to even out the power distribution from an engine. In early steam engines, such flywheels were large and heavy and would spin slowly—tens of hertz, or a few hundred revolutions per minute. Modern flywheels are much smaller and lighter and store a lot of energy by virtue of the fact that they spin much faster—up to 1,000 Hz, or

60,000 rpm. (The energy stored depends on flywheel mass and on the square of rotation speed, so speed counts for more.)[20]

There are two areas where high-speed flywheels are becoming the dominant means of storing energy. First, they are very useful in systems that require an uninterruptible power supply. The consequences of an interruption in power for such systems would be catastrophic. For example, computer data storage systems (for banks, Internet servers, the telecommunications industry, security systems, or airports) make use of flywheel energy storage because it responds very quickly, providing ride-through power until backup generators kick in. Flywheels are almost as efficient as batteries—typically 80%—and take up a lot less space for the same stored energy (i.e., they have greater energy density). One advantage of flywheels is that they do not degrade when power is drawn from them many times in rapid succession or when subjected to a high-power load that drains most of the stored energy in a few seconds. They have longer lives than do batteries and require less maintenance.

The second application for flywheels is in the automobile industry, where they are used for short-term energy storage. *Regenerative braking* is a technique that retains some of the kinetic energy of a moving vehicle, transferring it to a high-speed flywheel instead of dissipating it as heat, as happens in conventional brakes. The flywheel energy is then discharged when the vehicle accelerates. Thus, flywheels assist with saving energy and also wear and tear on the brakes.

Flywheels spin faster when energy is input and slow down when it is drawn off. They act like electric motors when electrical energy is supplied to them, spinning up to higher speeds, and like generators when energy is discharged. In automobile applications the changeover from energy input to output is achieved by changing the transmission ratio in a continuously variable transmission. The input and output rates can be very different. In other words, a small amount of input power applied for a long time can later be discharged rapidly as a single burst of great power. A small modern flywheel storage system can typically store a kilowatt-hour of energy, which may be released in many small bursts of power, or as a 100-kW burst lasting only a few seconds. This capability makes flywheel energy storage attractive for starting up engines and other equipment. At the very high rotation rates that can be obtained by modern flywheels, the wheel is subjected to enormous forces and so must be made very strong. The manufacturer must also produce zero imbalance—in other words, the wheel must be perfectly sym-

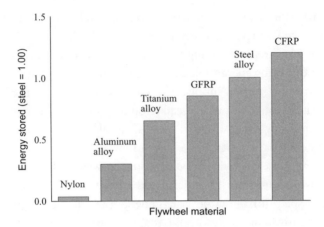

FIG. 1.8. Modern high-speed flywheel energy storage capability, for different flywheel materials. Here, I assume a solid cylindrical flywheel with a radius of 10 cm and a length of 20 cm made of various materials (GFRP = glass-fiber-reinforced polymer, CFRP = carbon-fiber-reinforced polymer). These materials are compared with the energy storage of a steel cylinder, which typically for a wheel of this size might hold 1 kWh of energy. CFRP beats steel, despite being lighter in weight, because it is stronger and so can be spun faster.

metrical about the rotation axis. The strong forces impose a maximum rotation speed upon the wheel; anything greater and it may fly apart explosively. In figure 1.8, I compare the energy that can be stored in flywheels that are all the same dimension but made from different materials.[21]

Alternative Energy Storage and Alternatives to Energy Storage

There are many other ways in which energy is stored, or redistributed. A key feature here and in later chapters is the notion, already mentioned, of *energy density*. Thus, a flywheel can store energy up to a maximum that is less than 1 kWh/kg, whereas a kilogram of gasoline contains about 14 kWh and a kilogram of hydrogen about 38 kWh. We will be investigating gas (and oil) in chapter 5, and so I will delay further discussion of these important energy sources until then. Clearly, however, the energy density figure alone is sufficient to explain why we use internal combustion engines and not flywheels to power automobiles. The even greater energy density of hydrogen makes it a very attractive substance for storing energy, and one that is being increasingly exploited. Hydrogen gas is generated by electrolysis of water.[22]

Later, the energy stored in hydrogen can be converted back into electricity via an internal-combustion-engine generator or via an electrolytic fuel cell; the process is 50%–60% efficient.

There are thermal energy storage devices. The most familiar example actually isn't, but I will mention it anyway. Ice can be made by an air conditioner at night, when electricity is cheaper than during daytime. The next day, air is cooled by being blown over the ice. (Technical quibble: heat is not stored in the ice but is removed from water which consequently freezes.) Energy can be stored in magnetic fields. A large amount of energy can be stored for a short period in superconducting magnetic energy storage devices by passing direct current along a superconducting coil. The refrigeration is expensive although the efficiency is high, at about 95%. The high efficiency and quick discharge time make magnetic energy storage attractive for systems that require high-quality power, and lots of it.

We have seen that energy storage is economical when the marginal cost (the difference between the price of electricity at night and during the day) varies more than the cost of storing energy and retrieving it when needed. The marginal costs vary with generator technology. Thus, base-load power plants (coal, nuclear) have low fuel costs but high maintenance and capital costs. If energy storage is insufficient for matching power supply to demand, or if it is uneconomical, then the alternative is a load-following power plant. These generating plants alter their output to match demand. As indicated earlier, this is not generally possible, or cost-effective, but some technologies permit it; gas-turbine generators are the prime example. Load-following power plants are characterized by low construction, operating, and maintenance costs but relatively high fuel costs—the opposite of base-load plants.

Energy is the ability to do work; power is the flow of energy—the rate at which it changes in time. Energy can take on many different forms and can be converted from one form to another, with varying degrees of efficiency. The cost and efficiency of power distribution depends upon the type of energy that is flowing. For national power grids, it is economical to adjust power supply to meet fluctuating demand. Such adjustments can be made by energy storage or by adjusting the power supplied. Many types of energy storage devices have found application in different sectors of industry.

2

WHAT ALL THE WORLD WANTS

In the early days of the first Industrial Revolution in England at the end of the eighteenth century—a time when mankind was on the cusp of changing the world—a far-sighted and savvy businessman liked to show visitors around his new Birmingham factory. Matthew Boulton possessed a gambler's instinct, and he bet heavily on a young Scottish engineer called James Watt. The factory was that of Boulton and Watt, a manufacturer of steam engines. Indeed, it was largely the success of Watt's development that gave rise to the technological takeoff that we label the Industrial Revolution, as we will see. Boulton is said to have told his visitors, "We sell here, sir, what all the world wants: power."[1]

Power, its various sources, and the history of how humans have developed it is the subject of this chapter. We begin at the beginning of our species, continue with a snapshot of our power generation capabilities today, and end with a puzzled glance at the future.

Power of Old

In prehistory we lived like other animals, and the power we needed for life was acquired from other animals and from plants. A million years ago the only power that the predecessors of *Homo sapiens* consumed was that contained in the food they ate. This power came, ultimately, from the sun.[2]

We have seen that a small fraction of the sun's power bathes the surface of the earth. A small fraction of this small fraction is taken up by photosynthetic plants, which convert the light energy into chemical energy. Herbivores eat the plants; carnivores eat the herbivores; omnivores like you and me eat both. An important point grabs our attention immediately, a point that will shoulder its way front and center throughout this chapter and the rest of the book. Energy density is almost as important as energy. How does energy density figure into the food chain? Prominently. The density of vegetation depends on sunlight: equatorial lands contain more edible foliage per square kilometer than does tundra. The number of herbivores per square kilometer is similarly distributed; there are more wildebeest per unit area of African savanna than there are moose per unit area of Canadian arctic. So it goes up the food chain: the area density of carnivores is limited by herbivore numbers.[3]

Fast forward to 100,000 years ago. As a species, we improved our food-gathering techniques and learned to harness fire, to keep ourselves warm and to cook our food. (Cooked food requires less energy to digest, freeing up more energy for our expanding brains.) The ability to control fire meant that the amount of energy consumed by each human per day—in essence, our per capita power consumption—increased. Thus, during this period, our energy was obtained from wood as well as from food. At the beginning of recorded history, about 3000 BCE, we were developing agriculture and a more sedentary and urban lifestyle. We made use of draft animals to plow our fields and cart our produce to market—add draft animals to our growing sources of energy and power.[4]

By the Middle Ages, say 1400 CE, humans had developed the first machinery to generate and distribute power. I will discuss the waterwheel and its more powerful if less reliable cousin, the windmill, in the next two sections. Half a millennium later, when much of the Western world, at least, had industrialized, we developed the steam engine fueled by coal; this fossil fuel is still a major source of power. (Note in passing the transformation of energy that led to coal deposits: nuclear energy in the sun radiated as light energy, which reached the earth and was converted by photosynthesis into chemical energy.) I will discuss the steam engine later in this chapter; coal deserves a chapter of its own.

Thus, by 1875 of the present era, mankind was generating energy and power from food, wood, draft animals, water, wind, and coal. Today we can add oil, natural gas, and the nuclei of certain atoms. These last four sources now account for most of the power we consume. In figure 2.1, I summarize the development of power sources throughout human history, showing the power consumed per person at different times in history and the sources of this power. Note that a man living in the developed world of 1970 consumed three times as much power as his ancestor a century earlier, and that this nineteenth-century man consumed three times the power of someone who lived in medieval times.

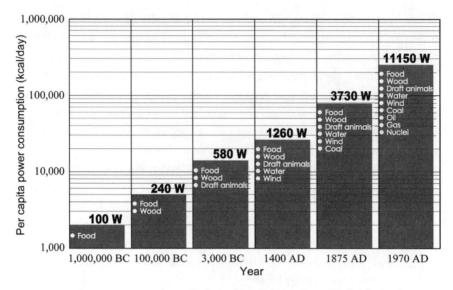

FIG. 2.1. Per capita power consumption throughout history. This power includes the power of machines that operate within our society, such as power plants. Note the logarithmic vertical scale. Power is expressed in both kilocalories per day and in watts; power sources are indicated in white text. *Graph based on data in Cook (1971) and Mattick, Williams, and Allenby (2009).*

Waterwheels

The utilization of power, the development of agriculture, and the use of fire are considered to be the most momentous inventions in human history. If we focus on a single millennium and a single *prime mover*, or power source, it has been claimed that the most outstanding technological development over the period 500 CE to 1500 CE was the waterwheel.[5]

The earliest waterwheel, the *Norse wheel* or horizontal wheel (so called because its axle was vertical), gave way in the classical period, when Greece and then Rome were flourishing, to the more powerful and familiar vertical wheel (which rotated about a horizontal axle). Neither Greeks nor Romans wrote much about waterwheels, and so it used to be thought that these wheels were not common until the Middle Ages, when they flourished. Nowadays historians are not so sure: plenty of archeological evidence indicates that waterwheel use was widespread—on virtually every Roman manorial estate, for example. There is also uncertainty about the place of origin of waterwheels. It is quite possible that they were developed independently in different parts of the world.

Whenever and wherever waterwheels were invented, they spread quickly across Eurasia, and for at least 1,500 years they were the most significant man-made source of power. A famous example is the complex in Barbégal, near Arles, France, a Roman "power plant" that operated for most of the second and third centuries CE. This plant consisted of two series of eight waterwheels on the side of a hill. Water would turn the top wheel and then tumble downhill to enter the second wheel, setting it turning, and so on. I estimate that the plant generated about 20 kW, in modern units. This power was used to grind wheat (27 tons of grain per day) to make flour.

Waterwheels were the main source of power at the start of the Industrial Revolution (the early factory phase, over the period 1780–1820) until steam engines took over. Between the Barbégal heyday and the Industrial Revolu-

---

ROCK CONCERT

To provide numbers for early power sources, let me transport you to Neolithic or Bronze Age England and put you to work raising a stone lintel to complete the construction at Stonehenge. Each of these stones weighed about four tons and needed to be raised about 4 m. Could you and your crew lift it? We'll say that the circumference of the stone is 8 m so that 13 men can crowd around it and get both hands under it. Each man would have to lift 300 kg, or 700 pounds: no, men can't have lifted these lintels. (Insufficient energy density, please note: not enough men can crowd around the stone to lift it.)

Perhaps you build an incline, place the lintel on rollers (tree trunks), and drag it up the 4 m height. Assuming reasonable values for the incline slope and for rolling friction, and assuming a power expenditure of 75 W per man, your crew could move the lintel up the slope in 5–10 minutes. If you had a block and tackle, you could have saved the effort of building an incline and lifted the lintel in 2–3 minutes. If you used two horses, you could discharge your crew.*

*A horsepower is 746 W and so replaces 10 of your men. However, real horses in olden times were less powerful than this modern measure. James Watt defined the horsepower unit for us (ironically, given that his steam engine put a lot of horses out of work): folklore has it that he arbitrarily increased the measured power of a horse by 50% to give us the definition (see Encarta 2005). Nowadays, horses are large enough to generate, appropriately enough, about 1 horsepower; see Stevenson and Wassersug (1993).

tion lay 15 centuries of increasing waterwheel use and application. The total power generated by waterwheels in Europe increased exponentially during the Dark Ages, and from about 1100 CE, European civilization became the first to be powered by something other than human muscles (draft animals and waterwheels, and then windmills). As well as grinding grain for flour, waterwheels were used for fulling cloth, blowing bellows, sawing wood, and (following the introduction of the crank and trip hammer from China, in the Middle Ages) for forging iron. In the early phase of the Industrial Revolution, they were also used for threshing and for spinning and weaving.[6]

The vertical waterwheel is either *undershot* or *overshot*, depending on whether the water that causes the wheel to turn passes under or over the axle. Undershot wheels (two historical examples are shown in fig. 2.2) were much more common because they required only that the vanes—the outermost part of the wheel—be dipped in water, whereas overshot wheels required that the water be applied at the top of the wheel. Thus, undershot wheels could be used in shallow, sluggish streams on flat plains, whereas overshot wheels were largely confined to hilly terrain or else required extensive infrastructure such as aqueducts to pipe the water to them (as at Barbégal). Overshot wheels were more powerful and more efficient.

Much effort went into developing a theoretical understanding of waterwheel engineering and physics during the eighteenth and nineteenth centuries, with the aim (successfully, as it turned out) of increasing wheel efficiency and thus increasing power. John Smeaton, in eighteenth-century England, tinkered with many machines of his day, including not only the waterwheel but also the windmill and the early "atmospheric" steam engine, and empirically improved them all. He showed that in practice, overshot wheel efficiency peaked at about 60%, whereas undershot wheel efficiency was about half as much—that is, an undershot wheel would convert about 30% of the water power into useful wheel power. (For more on the power of the overshot wheel, see the appendix.) Also in England at that time, Thomas Mead invented a mechanical device, the centrifugal or flyball governor, that helped to regulate the grinding action of a watermill; this device would find very fruitful application in Watts's steam engines.[7] Jean-Victor Poncelet, in nineteenth-century France, contributed much theoretical knowledge to undershot wheel design. His work anticipated the action of turbines, the successor of waterwheels.

FIG. 2.2. Waterwheels. (a) A twelfth-century Belgian undershot waterwheel. (b) Woodcut of a medieval European waterwheel used to raise ore from a mine. *Photo courtesy of Jean-Pol Grandmont; illustration from Agricola (Georg Bauer),* De re metallica *(1566).*

WHEEL POWER

It is easy to estimate the power generated by a waterwheel of given size if we know its efficiency. Let's assume that you have completed Stonehenge and are now working as a miller in medieval Europe. Your overshot wheel has a radius of 1 m. The *headrace* (the sluice or flume that directs river water to the top of the wheel) has a cross-sectional area of 2 m², and the water that flows along it moves at 0.5 m/s. Efficiency is, let us say, 30% (it will double over the next few centuries, as we have seen). A back-of-the-envelope calculation shows that your waterwheel generates about 6 kW (see the appendix). Many historical overshot wheels were much larger than yours, operated in faster-flowing water, and so could generate much more power (perhaps 50 kW), but your mill is much closer to the average.

Windmills

Windmills existed in Islamic lands before they reached (or, just as plausibly, were independently invented in) Christian Europe. Windy Afghanistan and Persia were dotted with windmills from the ninth century CE. These were vertical-shaft mills, with vertical sails that rotated around the shaft like sheets on a rotary washing line. In contrast, European windmills had horizontal shafts, and did not appear until about 1150 CE, in the windier parts of Europe—northern Germany, Holland, and southeastern England. One disadvantage of the European design was that the sails had to be turned to the wind, which could change direction unpredictably. (Vertical-shaft windmills do not have this problem.) So from their earliest days, European windmills had to be designed so that the miller could rotate the whole structure about a central post so as to face the wind. Such *post mills* were superseded in the sixteenth century by *smock mills*, in which only the cap at the top of the mill, and the sails, rotated; the base of the mill was fixed and covered with a wooden exterior. Smock mills, in turn, gave way to the epitome of traditional windmill design, *tower mills* (or turret mills); in the latter, the wooden smock is replaced by a stone or brick body, atop which the cap and sail are free to rotate. A tower mill, and the complex gearing that is required to turn the cap, are shown in figure 2.3. When we think of an old-fashioned windmill, say in the polders of Holland, we see in our mind's eye a tower mill.

Over the centuries, windmills became more and more sophisticated and consequently more and more powerful. Smock and tower mills were not turned into the wind by hand; rather, wind action on the *fantail* turned them automatically (see fig. 2.3c). Windmills grew taller, which increased their power, for two reasons. First, wind power depends upon the area covered by the rotating sails; a taller windmill could have longer sails.[8] Second, wind higher up off the ground is faster than wind lower down; this *wind profile* (an increase of speed with height above the ground) arises because of friction at the surface. Wind profile explains the odd angle of the wind shaft or axle: it was not quite horizontal but was instead inclined at an angle of about 10°. This angle meant that the sails would catch the wind better, given a wind speed that increased with altitude.

Size matters with windmills—size and sophistication. Winds are notoriously fickle and change quickly in both speed and direction. Wind speeds that are too high can damage a windmill sail. Consequently, sails became adjustable: their pitch (angle to the wind) could be changed and baffles like Venetian blinds could be turned to spill excess wind. Light winds, on the other hand, might not be able to turn the sails. This is where size came into consideration. In the appendix, I show that simple physical considerations predict that the power of a packet of wind passing through windmill sails increases quickly with increasing sail size and increases even more quickly with wind speed. (More precisely, wind power increases as the square of sail length and as the cube of wind speed.) Thus, larger sails could take advantage of lighter winds.

Since light winds occur more frequently than strong winds, large windmills could operate for more hours per year than could small windmills. For example, early Dutch mills (and presumably also other contemporaneous European windmills) could not operate in wind speeds below 12 mph or above about 22 mph, which restricted them to 2,700 hours per year—around 7 hours per day, on average. Later (larger) mills could operate in winds as slow as 9 mph and so could operate for 4,400 hours per year (say 12 hours per day). In Holland, the extra work mattered greatly because the Dutch used windmills to pump water out of land that had been reclaimed from the sea.[9] These pumps needed to be operating continuously or nearly continuously.

Windmills reached their peak of efficiency in the early twentieth century. This peak was never high: perhaps 5%. Even so, the power generated by a large European windmill—around 40 kW—exceeded that of most large waterwheels.[10]

These old windmills differed in purpose as well as design from modern wind turbines. The windmills of old were designed to act locally for one

FIG. 2.3. Windmills. (a) A traditional Dutch tower mill in the outskirts of Amsterdam. (b) The interior of Pantigo windmill on Long Island, New York, looking up into the cap. (c) A modern American windmill (this one is in New Mexico); note the fantail. The steel blades that were introduced in the 1870s proved to be very useful for small windmills. *(a) Photo courtesy of Massimo Catarinella; (b) photo courtesy of the Historic American Buildings Survey, Library of Congress, Washington, D.C.; (c) photo by author.*

specific purpose, be it grinding grain or pumping water. Modern wind turbines (of which, more in chapter 8) act in concert with dozens or hundreds at the same site to generate electricity. The generated electrical power is then added to a power grid and distributed far and wide.

Letting Off Steam

In the eighteenth and nineteenth centuries waterwheels and windmills provided Europe with much of its power. At their peak there were about 500,000 waterwheels and 200,000 windmills across Europe. Yet both these venerable machines fell from their summit, from abundant economic powerhouse to scarce and therefore protected tourist attraction, in less than a century, after twenty centuries (in the case of waterwheels; seven centuries for windmills) of continuous use and development. The reason for this relatively abrupt demise was, of course, James Watt's development of the low-pressure steam engine.

The steam engine is surprisingly old. In 1698 Thomas Savery produced a water pump that was powered by steam, and 14 years later, Thomas Newcomen produced his *atmospheric engine*, also used for pumping water (out of mines). Newcomen's engine represented the best technology in the field for six decades, until James Watt found a significant technical improvement.

Newcomen's engine made use of a single very large and slow-moving piston inside a cylinder. Steam was employed to power this engine, as follows:

- A boiler generated steam and introduced it into an elongated cylinder, which housed a piston.
- Water was then sprayed into the cylinder to cause the steam to condense, thus creating a partial vacuum and drawing the piston into the cylinder as a result of atmospheric pressure.
- Steam was reintroduced, restoring the pressure inside the cylinder and allowing the piston to move out, thus completing the cycle.[11]

The pressures involved throughout the piston cycle were never higher than atmospheric (hence the engine's name). Such low pressures were attainable in the early eighteenth century; higher pressures were not because manufacturing imperfections in the construction of iron and poor tolerances led to leakage. Nevertheless, the reciprocating action worked well enough as a water pump.

Watt's great idea was to introduce a separate condenser outside the cylinder. This design meant that the cylinder remained hot throughout the cycle and that only a relatively small compartment changed temperature, thus reducing wasted energy. Watt's idea dawned upon him in 1765; it took him until the year of the American Revolution before he could make it work in practice. Watt's team—which included a number of talented apprentices, such as William Murdoch, as well as the patient and moneyed industrialist Matthew Boulton—developed several mechanisms to improve the action of his reciprocating engine, rendering it more suitable for factory applications. *Sun-and-planet* gears and a crankshaft transformed the reciprocating motion of the piston into rotary motion; the *parallel-motion mechanism* led to a double-acting piston (with steam pressure on both in and out strokes of the piston cycle); the *flyball governor* regulated engine speed.

The result of all this tinkering was an engine which ran much more smoothly so that it could turn a shaft at constant speed. Constant speed was essential for industrial applications, and it was at this point (reached at the end of the eighteenth century) that steam engines supplanted waterwheels as the prime mover of industry. Indeed, because a steam engine could be housed anywhere (and not just beside a river), the application of steam power greatly increased the output of industry, which in turn increased the demand for Watt's steam engines.

John Smeaton had earlier tinkered with Newcomen's atmospheric engine to make it as efficient as possible. He found that a 40-hp Newcomen engine consumed 15.87 pounds of coal per horsepower. Later, a Watt engine of the same power consumed only 6.26 pounds of coal per horsepower. Not only was the Watt engine more efficient, but it was also smaller—it had increased energy (and power) density. Thus, a Watt engine could be used in a small space. Steam engines were applied to coal mines (which provided fuel for steam engines) and iron production (which provided the raw material for making steam engines), as well as to power looms and other factory machines. Power looms displaced hand-loom weavers, creating social unrest but increasing the production of cotton goods and improving their quality. The Industrial Revolution took off because of the steam engine,[12] and so the company of Boulton and Watt flourished.

Watt was by no means the only innovator whose ideas contributed to steam engine development. Indeed, Watt's engine was not the leading design. His engine was low-speed and low-pressure, akin to Newcomen's; the future lay with more powerful high-pressure engines that cycled at much higher speeds—engines that relied on steam pressure to push cylinders rather than on condensation and the resulting vacuum to pull them. Boulton and Watt held a patent on steam engines, however, and they actively suppressed innovations by others that threatened their own success. In 1800 Richard Trevithick developed a high-pressure steam engine that dispensed with the old condenser idea. Watt's former assistant William Murdoch came up with a high-pressure design, which was blocked by Watt. Two decades earlier, in 1781, Jonathan Hornblower had developed a compound (double-cylinder) engine[13] and was prosecuted successfully by Boulton and Watt for infringement. Hornblower died impoverished and disappointed—yet compound engines were the way of the future. The early history of steam engine evolution is both littered and enriched by the contributions of lesser-known individuals: Polzunov, Woolf, Bull, Cartwright, Evans, Cugnot, Griffiths, Gordon, Gurney, Hancock, and many others.[14]

A good idea cannot be suppressed forever: high-pressure engines found application in steam locomotives, with the work of Trevithick and of the father-and-son team of George and Robert Stephenson, in the first quarter of the nineteenth century. Steam trains (fig. 2.4) exploded across the world;[15] the number of miles of rail track across Britain, and then across Europe and the United States, expanded exponentially, permanently changing the face of these lands and the economics of the countries they covered.[16]

Further practical and theoretical developments in the third quarter of the nineteenth century increased the power and efficiency of steam engines. Jacketing of the cylinder, superheating the steam, increasing the steam inlet pressure, and perhaps most significantly, *compounding* the engine all contributed to this increase. Compounding meant introducing several stages to the expansion of the steam, a process that extracted more energy from it and so wasted less. A piston was powered at each stage—in terms of energy production, it was like a waterfall which fell in several stages—with energy extracted from each stage by a waterwheel (as at Barbégal), instead of falling once with a single waterwheel at the bottom. By smoothing the cyclic motion of the pistons, compounding also reduced wear.

The logical limit of the compounding idea is the steam turbine. In such a turbine, the steam flows continuously instead of being cycled, and the turbine blades convert steam energy directly into rotary motion without the intermediate stage of linear reciprocating motion. While steam engine efficiency rose over the decades from about 1% to 10%, steam turbine efficiency is 45%. Thus, steam turbines were, right from their inception in the 1880s,

FIG. 2.4. Railroads opened up the United States; they changed the nation and the world. This steam locomotive is from the Cumbres and Toltec Railroad in the American Southwest. *Photo courtesy of Dr. Charlotte Geier.*

much more efficient than earlier steam engines. The inventor of the steam turbine, Charles Parsons in England, lived to see his creation expand from its earliest application, in marine engines, to use in electrical power generation —its main role today.[17] By 1899 a single Parsons turbogenerator could produce 1 MW of electrical power. Today, 80% of electricity is generated using steam turbines.

The theory of steam engines was developed in stages by Carnot and Clapeyron in France and by Rankine in Britain. Turbine theory was developed by Stodola, a Slovak working in Switzerland. It is typical of the rapid rate of technological progress in the nineteenth century that in both cases— steam engines and steam turbines—theory lagged practice. That is to say, the theories of how these machines operated were worked out after the machines had been invented and during the period of their deployment and further development. Needless to say, many more people than I have indicated here were involved in both the practical construction and the theoretical understanding of turbines.[18]

Coal—A Dark History

Steam engines were powered by coal—and were used to mine coal, by pumping water out of the mines. Coal mining has historically been a dangerous trade, as we will see. Because coal was cheap and plentiful, it remained a power source, fueling factories and power generating stations, long after steam engines were relegated to history. Nowadays industrial power is derived from other sources, which we will investigate, and coal is mined almost exclusively for use in generating electrical power.

Coal is a rock that burns; it consists of the fossilized remains of plants from the Carboniferous period (360–300 million years ago) or from earlier algal deposits. Thus, it is biological in origin, and the chemical energy it contains derives ultimately from the sun. There are three main grades of coal and numerous subdivisions, all of which relate to the quality of burning. The lowest grade coal is *lignite*, or *brown coal*, and the highest is *anthracite*. In between there are the *bituminous* coals. Grade is generally correlated with depth of the deposit below ground level, with higher-grade coal being buried deeper.

Historically, coal was used as a fuel infrequently until the beginning of the second millennium, although there is some evidence of occasional use as far back as the fourth century CE in China. The country that first began burning coal in quantity was England. This may not surprise you, given that the Industrial Revolution began there, but in fact coal had become an important fuel in England 500 years earlier, in the thirteenth century. This much earlier transition from wood to coal occurred because an increase in population and the resultant deforestation of the island led to a shortage of fuel for heating and cooking. The first records of pollution caused by coal burning in England date from the 1300s. By the 1700s, it has been estimated, over 80% of the world's mined coal was dug out of English ground.[19]

The Industrial Revolution greatly increased the use of coal. Ironworks had been notorious for their consumption of wood fuel, but when coal was converted to coke, it proved to be a better fuel for the purpose.[20] Iron production was key to the technological takeoff of the Industrial Revolution in England, and so coal was mined extensively. By 1848 Britain produced more iron than the rest of the world put together.

Coal pollutes doubly: it scars the land where it is dug out of the ground, and it pollutes the air and land when burned. The grimy industrial cities of the nineteenth century were a testament to this, with their belching smoke-stacks, blackened buildings, and foul air—accompanied in many cases by severe health problems of the citizens (in particular, respiratory diseases). Given these huge problems, the fact that many countries persist with coal burning today points to the importance of this fuel.

Coal kills not just by polluting lungs, but even before it is extracted from the ground. Deaths in coal mines were huge wherever coal was dug. That is still true today, as we will see in chapter 4, but the number of people killed in earlier centuries was much greater (per ton of coal extracted). The worst mining disaster in European history occurred in the Courrières coal mine in northern France in 1906, when 1,099 miners died. The worst in U.S. history happened a year later in Monongah, West Virginia, when at least 362 miners were killed. Coal mining is so dangerous because, as with any underground mining operation, there are the risks of mine wall collapse and of flooding. But other risks associated with coal mining make it particularly dangerous. Poisonous gases such as hydrogen sulfide (commonly known as "rotten-egg gas" because of its smell) can be released. Other gases are even more dangerous because they have little or no odor: *choke damp* (carbon dioxide) kills by suffocation, and *white damp* (carbon monoxide) poisons.[21] Other released gases, such as methane, are explosive.

To illustrate the historical problems of mining coal—the deaths caused and the social ills generated—consider the Mines and Collieries Act, passed by the British Parliament in 1842. This act forbade the practice, previously common, of sending women and children (as well as men) down into the mines. The act followed a freak flood in an English mine in 1838 that killed 11 girls between the ages of 8 and 16, and 15 boys between the ages of 9 and 12.

I will have much more to say about coal and coal production in chapter 4. Today coal is still an important source of fuel worldwide, especially for emerging nations. Coal matters even for fully industrialized countries such as the United States, as you may judge from figures 2.5 and 2.6. These graphics show the relative contribution of various fuels to the national power budget over time (coal was king during the period 1885–1950) and the use for these fuels today. After World War II petroleum overtook coal as the main source of power, and it is to petroleum that we now turn.

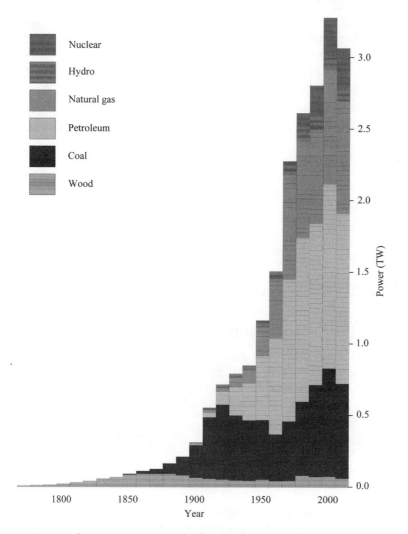

FIG. 2.5. U.S. annual power consumption by decade. The total is divided into fuel sources. This graph shows how the relative importance of different energy sources has changed over the years as well as showing how the total has grown. *Adapted from data provided by the U.S. Energy Information Administration (EIA 2011).*

The ICE Age

The internal combustion engine (or ICE) was invented in 1862 in Germany.[22] It is responsible for the widespread use of petroleum today, in the form of gasoline (petrol in England) and diesel. Petroleum is used mostly to

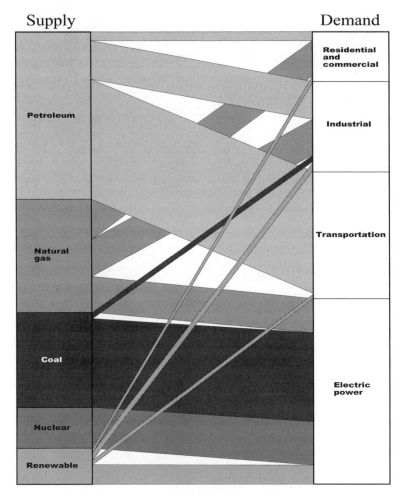

Supply   **Demand**

FIG. 2.6. Current U.S. energy sources and their allocation. *Adapted from data provided by the U.S. Energy Information Administration (EIA 2010a).*

provide fuel for transportation (fig. 2.6). Just as coal powered the first Industrial Revolution in Britain, so petroleum can be regarded as the fuel that powered the second Industrial Revolution in Germany, much of the rest of Europe, and the United States.

Petroleum, like coal, is a fossil fuel, with many of the same environmental and health issues. However, the geological distribution of petroleum is very different from that of coal, and so the geopolitical consequences are quite unlike. Why bother with petroleum? Why not use coal to power mechanical transport? Coal worked well for steam trains but does not work well for automobiles. The reason lies in the energy density of the two fuels. As measured by both weight and volume, the energy density of petroleum is almost twice that of bituminous coal. Since a mechanical means of transport must carry its own fuel, energy density matters. Large vehicles such as ships and locomotives managed quite well with coal, but the much smaller automobiles cannot carry enough coal to take them very far.

In chapter 5 I analyze the importance of oil to the present and future world energy budget. Here, I outline the historical development of oil and the oil industry. It goes hand in hand with the development and influence of the automobile. As is clear in figure 2.6, oil would be much less significant without a transportation infrastructure powered by petroleum. I will not spend time here on the development of the internal combustion engine,[23] but to set the table I should spend a few words emphasizing the impact of automobiles upon our lives.

The first modern cars were built in France and Germany in the 1890s, but automobiles for the common man were a product of early twentieth-century America. Why the United States? Because America had relatively inexpensive gasoline and Americans had relatively large disposable incomes. Most importantly, American carmakers turned the production of automobiles from a craft industry into the first modern production line. The famous Ford Model T rolled off such a line in Detroit in 1908; the almost equally famous moving assembly line that produced it began operation in 1913. By the time production was discontinued in 1927, over 15 million of these cars had been made and sold. By 1924 the manufacturing cost per car was down to $290, and the time to make one Model T was 1 hour 33 minutes. The methods developed for automobile manufacture changed factory production in general, as industrialists learned from the car producers.

The widespread use of automobiles called for and got a change in the infrastructure of the developed nations of the world. Paved highways were needed; these changed the look of our cities and countryside and created many jobs. Longer commutes became feasible, so suburbs sprang up around cities. Motels were invented, and malls arose on the outskirts of towns—all because of the increased mobility of people made possible by cars. There were more widespread economic consequences, directly connected to automobile manufacture and indirectly arising from support services. In addition to factories where cars and trucks were manufactured, the industry required sales outlets, repair shops, fuel distribution and delivery companies, and insurance. Transportation infrastructure became so important to the economies of nations that entire governmental departments devoted to highway development and maintenance arose.[24]

In 2010 about 78 million cars and commercial vehicles were being manufactured, compared with 58 million a decade earlier. More of these were made in China than in any other country—an indicator of the changing balance of economic power in the world—and more were made by Toyota than by any other company. (GM regained the lead in 2011.) Worldwide, there are over 800 million cars on the road.

No gain without pain: car culture comes with a cost. In many of the world's major cities (Ankara, Buenos Aires, Los Angeles, Mexico City, Milan, Tokyo, . . .) the air is barely breathable and leads to respiratory illnesses. Traffic congestion is another major headache around the urban centers of the world, and motor vehicle accidents are a leading cause of death and injury. In 1926 there were 23,000 fatalities caused by motor vehicle accidents in the United States—a quarter of all accidental fatalities that year. Today, there are typically 43,000 deaths per year in America due to car accidents. This is three times the number of homicides and represents 14 deaths per 100,000 of the population per year. (In Western Europe, there are 30,000 deaths per year due to automobile accidents, or 7 deaths per 100,000 of the population per year.)[25]

Oil has been used by man since antiquity. Natural oil seeps were exploited in the Middle East and in China, on a small scale, for lamp fuel. Oil also existed as natural pitch lakes; one in Trinidad was discovered by Sir Walter Raleigh, who used some of the pitch to caulk the wooden hull of his sailing ship. The lake was exploited for this purpose throughout the Age of Sail.

The first commercial oil refinery was at Baku, the capital of Azerbaijan on the western shore of the Caspian Sea. In 1861 Baku refined 90% of the world's oil, which was used to make kerosene for oil lamps. The rapid increase in motor car production led directly to a rapid expansion of oil extraction and refining throughout the world. By 1910 oil was being drilled in western Canada, Mexico, Persia, Peru, Russia, Sumatra, the United States, and Venezuela. The discovery of oil in a number of places in the Middle East in the 1930s was a key event in world geopolitics. Today, 80% of the world's accessible reserves are underneath that turbulent part of the globe. OPEC came into being in 1960; during that decade, oil wells sprang up on the north shore of Alaska and in the North Sea. Oil extraction and consumption have expanded each decade since the beginning of the twentieth century; oil overtook coal as the most important source of fuel in the late 1950s.

Today, some 40% of U.S. power is derived from oil (see fig. 2.6). America consumes far more oil than any other country, at 18.8 million barrels per day. China is a distant second (but closing rapidly), at 8.3 million barrels per day; this number will be out of date by the time you read it. Consumption over the world has increased linearly from 56 million barrels per day in 1983 to a little over 87 million barrels per day in 2010. Our dependence on oil became very apparent during the economic crises of the 1970s; from that time people began to realize that the supply was limited by nature, as well as by OPEC, and that one day this supply would be exhausted. Much more on this debate in chapter 5.[26]

Natural Gas

Natural gas, when it comes out of the ground, is a mixture of hydrocarbons. Typically, the content is 70%–90% methane, plus 0%–20% ethane, butane, and propane, plus a small amount of non-hydrocarbons such as sulfur dioxide and carbon dioxide. Many natural gas deposits are associated with oil wells. In fact, in the late nineteenth and early twentieth centuries, the natural gas that emerged at wellheads was considered a nuisance and was burned off (what a waste!). Some natural gas deposits are "nonassociated," meaning that they are located where there appears to be no oil deposits.[27]

Gas in the ground percolates upward from deeper within the earth. Much gas is able to vent to the atmosphere. (Such natural gas vents have been known for thousands of years and in some cases have been exploited.) However, a lot of gas is unable to reach the surface because it is trapped within ice or because it is beneath a cap of impermeable rock. There are two basic theories concerning the origin of natural gas. Shallow deposits—say, between 1 and 6 km below the surface—are found in organic sediments; geologists assume that long-dead organic matter has been compressed by the weight of rock above or that it has been decomposed by underground microorganisms. Deeper deposits may not have an organic origin; they are assumed to have arisen from chemical reactions between elemental carbon and hydrogen-rich compounds, in the absence of oxygen.

Whatever the origin, the gas that emerges out of the ground is described as *wet* or *sour*. This gas is then *sweetened* to remove all the components except methane. (Some of the other hydrocarbon components, such as propane and butane, are of significant commercial value. The non-hydrocarbons are not —indeed, sulfur dioxide is a foul-smelling poison.) The sweetened gas, called *dry* natural gas, is odorless, colorless, nontoxic, and noncorrosive.

Wet natural gas was considered a nuisance in the early days of oil extraction because there were no facilities for sweetening the gas and there were few pipelines. Pipes were expensive, as was the sweetening process, so natural gas was uneconomical except in the immediate vicinity of the wellhead. The first commercial use was for street lighting. Then in 1885 a German chemist named Robert Bunsen invented a burner that mixed gas with oxygen to produce a safe source of heat. Since that time, natural gas has been used for domestic heating and cooking. (It was advertised as "smudge-free heat" to emphasize its relatively clean nature compared with other fossil fuels.) The use of natural gas expanded greatly after World War II; with improved metallurgy and welding came less expensive pipelines, and it became profitable for oil companies to distribute gas. Later, the gas industry learned how to liquefy the sweetened gas, reducing its volume by a factor of 600, so that it could be transported overseas in bulk containers (at a low temperature of $-160°C$, or $-260°F$).

Natural gas has a mass energy density that is about 10% greater than that of gasoline, but its volume energy density is about 70% of that of gasoline. It has a higher energy density than coal and is much cleaner to burn. It is also very abundant around the world. The largest existing gas field is beneath Iran and Qatar, although a newly tapped field on the Louisiana-Texas border may be larger.[28] The biggest proven reserves of natural gas are in Russia. It is predicted that natural gas will form an increased fraction of the fuel used to generate electricity in the future. (Its consumption is currently split three ways equally: electrical power, industrial, and residential and commercial, as you can see in fig. 2.6.) It constitutes a quarter of the world's energy resources; the United States may have enough to last a hundred years (at the current rate of consumption—an important qualification). Today, more than half of Americans rely on natural gas for their domestic heating fuel.

Hydro History

We saw earlier that in the nineteenth century, waterwheel design gave rise to turbines. Starting in the early 1880s, technological advances and a rapidly increasing demand for electricity combined to propel *hydropower*—the application to electricity generation of water-driven turbines placed inside dams—to a position of prominence within a few decades.

One of the first instances of water power being used to generate electricity occurred in Grand Rapids, Michigan, where water-powered dynamos provided storefront lighting. In Niagara Falls, New York, street lighting was powered by water. The world's first hydroelectric plant was completed in 1882, in Appleton, Wisconsin; this was a 12.5-kW facility on the Fox River. Before the end of that decade, 200 locales in the United States were using water power to generate some if not all of their power. By 1907 hydropower provided 15% of U.S. electricity generation; by 1920 that figure had leaped to 25%.

A big boost to hydropower projects came as a response to the Great Depression, when the federal government built a number of large dams. (Other countries, notably the Soviet Union, undertook similar schemes during this period.) The idea was partly to create jobs and partly to increase the national electricity generation capacity. Among the dams that were constructed at this time were the famous Hoover Dam (on the Colorado River bordering Arizona and Nevada) and the Grand Coulee Dam (on the Columbia River in Washington), both of which are still an important part of U.S. hydropower today. Figure 2.7 gives some indication of the scale of the turbines employed in these dams.

By 1940 hydropower was providing 40% of America's electricity. The figure has become much less in succeeding decades, not because of any reduction in hydro generating capacity—on the contrary, it has increased greatly since 1940—but because other sources of electricity have increased even faster. In 2003 the United States had an 80-GW hydroelectric capacity plus a further 18 GW of pumped storage, and this represented less than 10% of total generating capacity.

Other countries developed hydropower according to their means and their geology: the use of water power to generate electricity requires large rivers and valleys that can be dammed to build up the *head* (height) of water. Not all areas of the world are suitable, so the distribution of hydropower internationally is patchy. Brazil, the Democratic Republic of the Congo, Norway, and Paraguay, all countries with a fortunate geography of big rivers or high rivers, each generate at least 85% of their electricity via hydropower. The top five countries in annual hydropower production are currently (now that's almost a pun) China (652 terawatt-hours), Canada (370 TWh), Brazil (364 TWh), the United States (251 TWh), and Russia (167 TWh). Note that

FIG. 2.7. Grand Coulee Dam turbine parts. (a) Francis turbine runner; (b) Francis turbine inlet scroll. A link to earlier waterwheel technology, the curved rotor blade shape was invented by James Francis in the 1850s. As we will see in chapter 6, the high efficiency of turbines depends sensitively on the shape of the turbine rotors and the water inlet. *U.S. Bureau of Reclamation photos.*

THE HEIGHT OF POWER

My calculations on overshot waterwheel power can readily be applied to modern hydroelectric dams so long as we increase the efficiency to 90% (turbines convert about nine-tenths of the water power into electrical power). We can substitute numbers for, say, the world's largest hydropower plant, the Three Gorges Dam in China. It has a maximum hydraulic head (water height difference) of 113 m. Power is generated by 29 Francis turbines, each of which is designed to operate with a water flow rate of between 600 and 950 m^3/s, resulting in a total generating capacity of between 17.3 GW and 27.5 GW. In practice, the maximum power of the dam is rated at 22.5 GW, so you can see that a rough calculation produces a good ballpark figure.

I have switched from units of power to units of energy. This is to emphasize a point. A large volume of flowing water generates a lot of power but not during summer if it dries up; energy figures indicate the yearly average flow of water and are perhaps a more useful indicator of capacity. Globally, hydropower contributes some 16% of installed capacity—about 890 GW of power supplying each year over 3,400 TWh of electrical energy.[29]

The Nuclear Age

In the introduction, I indicated that there would be little politics in this book. By that I meant I would not be expounding a personal political view but would instead be presenting the facts as I understand them, from the perspective of a physicist and engineer. But it is impossible to engage in the energy debate or to discuss certain aspects of power generation without appearing political. The reasons for this are probably clear to you: the matter is one of extreme importance to mankind, and whichever path we choose for the future will have huge economic and environmental consequences. Money, jobs, and public safety are at stake, in a big way. Emotions become engaged, and reason sometimes goes up the smokestack. Nowhere is this more apparent than in issues surrounding the subject of this section.

The world "nuclear" by the smallest transposition becomes "unclear," which does not bode well. The common mispronunciation of the word, "nucular"—even by U.S. presidents—doesn't help. Add to these quibbles the much more potent public introduction to the power of atomic nuclei, in

1945, and we have a recipe for misunderstanding and for strongly held misapprehensions. Thus, many nonscientists hear an initial explanation of the enormous energy contained within an atomic nucleus and then, on the heels of that, learn of Hiroshima and Nagasaki, and naturally associate nuclear power with atom bombs.

Nuclear energy is based on the *strong force*, a fundamental force of nature, along with the much more familiar gravitational force and electromagnetic force. This force is unfamiliar to most of us because it (along with the fourth force of nature, the weak nuclear force) is locked up inside atomic nuclei—or distant stars—most of the time. It can be conjured up and released as a very powerful explosion via unconstrained nuclear chain reactions. The prompt application of this knowledge produced the trinity of atomic bombs in World War II.[30] Less readily we learned to control these chain reactions so as to release the nuclear power more slowly, resulting, from 1954 on, in nuclear power plants. Typically, this power is used to heat up water to form steam, driving steam turbines to generate electricity.

Much more about the workings of nuclear power plants, and about the public disquiet and fear surrounding them, is provided in chapter 7. Here, I very briefly outline the historical development, but first a word about the politicization of the energy debate. Though this politicization (which all too often generates heat and not light, metaphorically) applies to all kinds of large-scale generation technology, the nuclear debate invokes particularly strong opinions and feelings.

To make a rational decision about which technology we should adopt for future power generation, we need reliable data. There is plenty of information out there; the trouble is that much of it is contradictory. Because of emotional attachment to one technique or, more commonly, disenchantment with another, some writers on the subject have presented us with the data that fit their viewpoint rather than the data that form it. At best such arguments present incomplete and misleading data; at worst the selection of data is disingenuous. Thus, one report has claimed recently that the total number of fatalities attributable to the Chernobyl nuclear disaster of 1986 will approach one million; Greenpeace claims 200,000; others claim that the number of fatalities was only 31 and that wind farms kill more people than do nuclear power plants.[31]

My approach to the plethora of data—to sift it and arrive at a believable core—is based upon long experience as a scientist, and upon wider commonsense notions. I find reports that conclude "*X* is better than *Y*" to be more believable than those that say "everything about *X* is good; everything about *Y* is bad." If power generation technology were so clear-cut, we would not have the range of generation approaches we see today: coal burning, oil and gas burning, hydropower, and nuclear power. Oil may be both dirty and dangerous, but it must have something going for it to be the fuel of choice for transport. Solar power may be clean and renewable, but, as we have seen, it cannot be a major player because it is expensive and (to date) inefficient.

We may be tempted to simply reject the most positive and the most negative conclusions as the consequence of mental distortions or plain lies from biased extremists. Thus, the figure that is (to my mind) closest to the truth concerning the total number of Chernobyl fatalities is about 4,000. There were 31 deaths within days of the accident due to massive radiation dosage; the remainder is determined from the known quantity and nature of leaked radiation and from the statistics of radiation epidemiology. (More on this subject in chapter 7.) But is it right to always reject the extremes?

Another area of confusion and misinformation concerns the apples and oranges of data that are placed before us, particularly concerning health and safety issues, which are often less clear-cut than, say, power output or cost per terawatt-hour. The data may have been selected to favor a particular viewpoint or may not be directly comparable with other data. The number of deaths in a power plant explosion may represent the whole story for an underground coal mine but not for a nuclear power station, as Chernobyl showed. The fatalities per million tons of coal dug out of the ground are very high if you read the figures for China but much lower if you are looking at data for the United States.

I present the data that I consider to be the most realistic and representative for each of the power industries and let the chips fall where they may. That is what I meant by "little politics."

The theory of nuclear fission—the breaking up of a heavy nucleus such as that of uranium and the consequent release of nuclear energy—was worked out in 1938 by a German-Austrian team. The breakup releases neutrons (subnuclear particles), which then collide with other nuclei and break them up, releasing more energy—a chain reaction. The first experimental chain reaction to achieve criticality (the release of enough neutrons to make the chain reaction self-sustaining) was produced in Chicago in December 1942 by a team of physicists under the direction of Enrico Fermi. This team and its work were subsumed into the Manhattan Project, the largest single technical program of World War II, which resulted in the world's first application of nuclear power: the obliteration of the cities of Hiroshima and Nagasaki in the summer of 1945.

The first application of controlled nuclear fission was also military: the propulsion of large surface ships and submarines. From the late 1940s, the United States, Great Britain, Canada, and the U.S.S.R. also pursued the peaceful harnessing of nuclear power. In 1947 the U.S. Atomic Energy Commission was formed. Under its auspices, and President Eisenhower's "Atoms for Peace" program of the early 1950s, the first U.S. nuclear power plant came online, near Shippingport, Pennsylvania, in 1957. (The same year saw the birth of the International Atomic Energy Agency and Euratom, the European Atomic Energy Commission.) The 60-MW Shippingport plant provided electricity for a quarter century before being closed in 1982. A 50-MW nuclear plant at Sellafield in the United Kingdom was commissioned in 1956 (it was later upgraded to 200 MW and remained in service until 2003). Unlike Shippingport, this facility was not intended entirely for peaceful purposes: as well as supplying electricity for civilian use, it produced plutonium for the military. The world's first nuclear power plant to generate electricity for a power grid was behind the Iron Curtain—a small (5-MW) facility at Obninsk in the Soviet Union that began operation in 1954.

It was also in 1954 that Lewis Strauss, chairman of the Atomic Energy Commission, famously said that one day, electricity supplied by nuclear power would be "too cheap to meter." Nearly six decades later that prophecy has hardly been realized, though it is likely that Strauss was taking a longer view and was considering another form of nuclear energy release, via *fusion* rather than fission. I say more about fusion—and about the different versions of fission reactors adopted by different countries—in chapter 7.

From the mid-1950s until the 1970s the nuclear program glowed (perhaps an unfortunate choice of words). Civilian nuclear power plants proliferated, and the power that they contributed to electricity production blossomed. World capacity grew from less than a gigawatt in 1960 to 100 GW in the late 1970s. Then followed a "brownout" as the number of nuclear power plants leveled off. Many dozens of proposed facilities were cancelled, some even at a late stage. The world's nuclear generating capacity has remained at around 16% of the total power supply since 1980. (The power generated by nuclear means has increased, to about 350 GW in 2005, but the share of the total has stagnated.) The reasons for this plateau are varied:

- reduced electricity demand during the recession of the mid-1970s;
- very long time overruns and significant budget overruns during the construction of nuclear plants;
- increasing regulatory requirements in the Western world, often resulting in costly retrofitting; and
- in the United States, the cost of litigation and growing public apprehension about the safety of nuclear power plants following active lobbying by environmentalist groups.

This last factor—public concern—grew following the well-publicized and costly accident at the Three-Mile Island nuclear facility near Harrisburg, Pennsylvania, in March 1979. This accident caused no fatalities, unlike the disaster at Chernobyl in the Ukraine seven years later.

There has been a revival of nuclear power from the turn of the century because of

- a large increase in electricity consumption by emerging nations;
- increased energy security awareness;
- a desire, ironically fueled by environmentalists (most of whom oppose nuclear power), to reduce carbon emissions; and
- the availability of "third-generation" nuclear reactors.

This brief resurgence has once again been put on hold following the nuclear catastrophe in Japan in 2011, when an earthquake gave rise to a tsunami that killed more than 15,000 people and caused the cooling system of the Fukushima-Daiichi nuclear power plant to fail. A large amount of radiation leaked out following the meltdown of three reactors, resulting in the evacuation of hundreds of square miles of the surrounding area. Again, I defer further discussion of these matters until chapter 7.[32]

Future History?

I have skimmed the surface of human power sources and their historical development. You will appreciate that a thorough analysis of these sources would require a thick book for each of them, so my summary has been necessarily terse. This historical approach has introduced us to our present mix of power sources—sliced and diced in various ways and presented in several graphs throughout this chapter—and naturally brings to mind several questions about the future. Can our current sources of energy and power last? For how long and at what cost, monetary and in other ways? What will the mix look like in 50 years' time? What are our long-term options?

In the remaining chapters of this book I provide a dispassionate view of the different technologies introduced here and leave you to form your own opinion about the best way forward. The last chapter provides my own views on the subject. The spread of options is remarkably broad, from nuclear fusion to waterwheels. (Yes, waterwheels are making a comeback, as we will see in chapter 6.)

Here are a couple of graphs to stimulate your neurons to think about energy and power generation. (Your brain, by the way, consumes fully 20% of the energy that you take in each day, though it constitutes only a little over 2% of your body weight.) In figure 2.8 you can see some of the imperatives for the United States of making the right choice for the future (or choices—a mix of technologies is likely to be safer than putting all our eggs into a single technological basket). Before 1958 the United States produced all the energy that it needed. Since then, Americans have had to import energy—an increasing amount and an increasing fraction of the total energy consumed, as shown. This perhaps awkward fact has geopolitical implications, in terms of energy security. In times of political turmoil, the ability of a country to run its factories and function economically depends upon a secure source of energy; a net importer is vulnerable to embargoes. A net exporter, on the

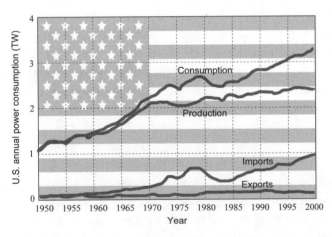

FIG. 2.8. U.S. power consumption and production (in terawatts) for the last half of the twentieth century, and imported and exported power for the same period. *Data from EIA (2010b).*

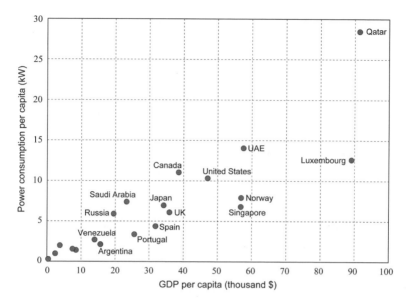

FIG. 2.9. Electrical power consumption per capita for 20 nations (in kilowatts) versus GDP per capita (in thousands of dollars). *Data from IEA (2008).*

other hand, may exert influence on another country or may find its own product boycotted. The reasons that a given country is a net importer or exporter are many and varied. Historically, America has been an economic powerhouse, and (North) Americans consume far more power per person than most other citizens of the planet; certain resources are distributed around the world unevenly—as we have seen in the case of hydro power, for example; certain countries are able to exist as independent entities precisely because they posses a large natural energy source—as in the Persian Gulf.

In figure 2.9 we see how electrical power consumption across the world is correlated with GDP. The correlation is not perfect (the data are scattered) because power consumption depends on many other factors, such as climate (more power is expended on air conditioning in Arizona and in the Persian Gulf states than in the United Kingdom, for example). Nevertheless, there is a clear trend: wealthy people consume more power than do less wealthy folk. What, if anything, does this tell us about power needs? We may be able to explain why Canada and the United States buck the trend (in that Canada has a lower GDP but consumes more power per person) based on winter fuel needs and distances traveled, but what about the difference between Saudi Arabia and Qatar?

Part of the answer lies in climate, as suggested: domestic consumption varies between about 15% and 40% of a nation's power budget. Part is economic and cultural: the relative importance of industry versus the commercial or public service sector varies significantly from one country to the next. Thus, the United States spends less of its power on industry than the world average (China spends more) and more on commercial and public services (China spends much less). The annual energy consumption of the United States has decreased by 2% since 1990 (for China, it has increased by 111%).

One final question: does the correlation between per capita GDP and power consumption imply a cause and effect? In particular, if we need to reduce our national energy consumption, will this lead to a reduction in GDP? Food for thought.

The power consumed per person has increased greatly over thousands of years; this increase has been particularly rapid in the last couple of centuries. Food was the original and only source; then we burned wood, then harnessed draft animals. Our first manufactured power sources, waterwheels and windmills, were used until the nineteenth century. Then coal, oil, natural gas, hydropower, and nuclear power became significant sources. In recent decades large numbers of people have become concerned about the sustainability of fossil fuels and about the safety of these and other power sources (especially nuclear), and have asked which source(s) we should concentrate upon developing and exploiting in the future.

3

THE VITAL SPARK

Electricity is the medium, not the message. It is the vehicle through which power is transferred very rapidly, and quite efficiently, from generating plant to user. We saw in chapter 1 that electrical power is not readily stored but that it moves quickly. Electric current is the common currency of power, readily converted as needed. Indeed, *current* and *currency* have the same etymology, referring to the idea of being in circulation. Thus, the gravitational power of water backed up behind a dam is converted into electric power by a turbine. Then, it is transmitted over lines from, say, Revelstoke in British Columbia to Spokane in Washington State,[1] where it is converted into mechanical power for Mrs. Ayes's tumble dryer, light for Mr. Bee's department store, and chemical energy for Dr. Cee's hybrid car battery.

This chapter is about the characteristics of electric power transmission. The subject is different from the main topic of the book—*sources* of power—but it is so central to any sensible discussion of energy sources and power generation that it must be included front and center.

Essential Electrodynamics

You will perhaps be glad to learn that I do not intend to provide a comprehensive account of this very complicated and mathematical subject, but there are one or two technical aspects of electrodynamics (the physics of electricity and magnetism) that lie at the core of electric power generation, transmission, and distribution, and these I do need to air. One issue is the complexity of the theory. It has hindered, historically, our understanding of the physics and development of power transmission technology. Thus, the genius of Edison did not extend to mathematical analysis; and as a consequence, he backed the wrong horse when pushing for a DC rather than AC electricity distribution system.[2] But I am getting ahead of myself; let me back up and paint a broad-brush picture of key elements pertaining to electro-dynamics.

Every material thing from aardvarks to zinc is made up of atoms; atoms themselves are made up of smaller constituent particles, like big fleas with little fleas on their backs to bite them.[3] An atomic nucleus consists of positively charged protons and electrically neutral neutrons. This nucleus is orbited by much lighter electrons, each of which has a negative electric charge. The number of electrons equals the number of protons, so that the atom is electrically neutral. What *is* electric charge? Too philosophical a question for this book (besides which, I don't have a cogent answer). You already have an intuitive idea of electric charge, and that will work well enough for our purposes. Opposite charges attract and like charges repel. This electromagnetic force is much stronger than the gravitational force. (You pick up a pencil; the force of your muscles is electromagnetic and easily overcomes the force of gravity of the entire earth acting on the pencil.)

Electricity is the movement of electrons within a *conductor*, a conductor being a material such as copper or aluminum inside which electrons are able to flow easily. Easily but not without resistance. Moving electrons clatter into atoms of conductor material, which impede their movement—this is electrical *resistance*. *Direct current* (DC) is the flow of electrons within a conductor. *Alternating current* (AC) is the oscillation of electrons back and forth within a conductor. The movement of electric charge carries power. This is not hard to imagine with a DC current, but if electrons are merely jiggling back and forth in an AC current without moving along the wire, then how is electrical power transmitted along the wire? Think of a water wave. The wave consists of water molecules that move up and down but do not move much along the wave direction (except in the case of tidal waves), and yet the wave carries energy with it: it has power.

One complicating aspect of electrodynamics that we will need later on is the idea of *induction*. An electric force field is created (induced) by a changing magnetic field, and a magnetic field is induced by a changing electric field. What have electricity and magnetism to do with each other? Everything; they are two sides of the same coin. This fundamental fact, plus the laws that underpin electrodynamics and allow us to quantitatively predict its effects, were first understood by the Scottish physicist James Clerk Maxwell in 1861.[4] Earlier in the century, important laws were discovered experimentally by others (Ohm, Joule, Faraday, Henry, Lenz, . . .), but it was Maxwell who brought them all together. The phenomenon of induction provides the basis for electricity generators, transformers, electric motors, and solenoids and gives rise to *reactance*—a form of electrical resistance that applies only to AC (not DC) currents.

In the remainder of this section I summarize the effects that the physics of electrodynamics has on the transmission of electric power. Some of the important practical consequences of induction are explained in the sidebars

---

INDUCTION AND ELECTROMAGNETS

We can develop a qualitative understanding of many electromagnetic effects if we know the basic idea of induction, a simple example of which is shown in the figure. If an electric charge moves, then *Faraday's law of induction* tells us that a magnetic field is generated and tells us how strong the field will be. The moving electrons within a current flowing along a wire generate a magnetic field around the wire, as shown in the figure. The direction of the magnetic field is determined by the direction of the current.

Now look at what happens when the wire is bent to form a coil: the magnetic field lines run through the inside of the coil. (They form a closed loop, but the field strength outside the coil is much weaker than inside because the field is spread throughout a much larger volume.) The coil acts like a bar magnet—indeed, the induction effect is enhanced by inserting an iron bar inside the coil, as shown. Which end of the bar forms the north pole and which the south pole depends upon the flow direction of the current. This construction is an electromagnet.

(a)   (b)

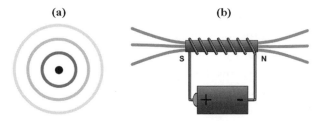

Induction and electromagnetism. (a) A wire (*black dot*) points out of the page. An electric current flows along it, generating a magnetic field (*circles*), according to the law of induction. If the current is moving out of the page, the magnetic field flows counterclockwise; if the current is into the page the field flows clockwise. (b) A wire which is coiled to form a loop is then connected to a battery, as shown. Electric current in the wire induces a magnetic field (*thick lines*) that is concentrated inside the coil. An iron bar enhances the effect. This construction is called an electromagnet. The north and south poles of the magnet are reversed if the battery is reversed—that is, if the current flow changes direction.

ELECTRIC MOTORS AND GENERATORS

There are many different types of electric motors, both AC and DC. In the three-phase AC induction motor shown schematically in the figure, three stationary electromagnets are arranged about a common center. (For clarity, the wires and batteries are not shown.) The AC currents are phased differently, such that the polarity of one of these *stators* is opposite to that of the other two.* In the figure, stator 1 is polarized differently from stators 2 and 3: its north pole is near the center of the motor. Because of the AC current, polarization of the stators flips back and

A three-phase AC induction motor. Stators 1, 2, and 3 are shown with their interior polarities configured north, south, and south (N, S, S). A small fraction of a second later (one-third of an AC cycle later), the stator polarities have changed to S, N, S, and then to S, S, N, and then finally returning to N, S, S. The cycle repeats for as long as AC current flows. In other words, the configuration shown rotates clockwise. Meanwhile, the N pole of the central rotor is attracted to the two nearby S poles of stators 2 and 3 (and its S pole is repelled). When the stator configuration rotates, so does the rotor, as its N pole tries to keep close to the S poles of the nearby stators. In this way, the axle of the electric motor (black circle) turns clockwise for as long as current flows.

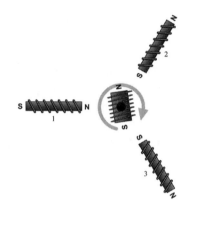

"Induction and Electromagnets" and "Electric Motors and Generators." I will concentrate on the long-distance transmission of power, as this is the main consideration for us.

First, physics tells us how much power is transmitted down a line when an electric current flows through it: power equals voltage difference (between one end of the line and the other) multiplied by current. Some of this power leaks out of the line and is wasted as heat. This leakage is a result of the resistance of the line to the flow of electricity though it. Resistance varies with the material: it is lower in conductors than in insulators, by definition, and is lower in copper (the usual conductor for wiring a house) than in aluminum (the usual conductor for making transmission lines because aluminum is cheaper and lighter than copper). The amount of power lost to resistance is determined by *Joule's law*,[5] one form of which states that this power is given by the square of electric current flowing through the wire forth. Thus, after a fraction of a second, stator 2 becomes the differently polarized electromagnet. Another fraction of a second later, stator 3 is the odd one. In this way the polarity of the stators rotates clockwise (in the case shown) at a rate determined by the AC current frequency.

Now consider the effect of this polarity rotation on the central electromagnet, which is free to rotate about an axis pointing out of the page. You can see that for the configuration shown, the central electromagnet (the *rotor*) will turn clockwise because its north pole will be attracted to the nearby south poles of the stators and will be moving to align its north pole between stators 2 and 3. Before it gets there, however, the stators' polarity will have changed, so that the rotor will turn to align its north pole between stators 3 and 1. Before it gets there, the stators' polarities have changed again—and so on and on. The rotor is continuously pulled clockwise, its axis turns—and we have an electric motor.

Electric motors are generators in reverse. Consider what happens when no current is supplied to the stators but, instead, the central axle is rotated by some external means (say the torque is applied by a turbine, a windmill, or an internal combustion engine on the end of a train of gears). The rotor movement induces current in the coils of the stators—and we have a three-phase generator.

* The phase of a signal (here an electrical wave) refers to the position of the wave compared to a reference position, at a given time, and is usually specified via a phase angle. If a 180° phase difference exists between two waves, one waves reaches a crest (a maximum) at the same time as the other reaches a trough (a minimum).

multiplied by resistance of the wire. To minimize loss of power due to electrical resistance, therefore, we should minimize the electrical current that flows; to maintain the desired level of power we must consequently increase the voltage. This is why power lines are always high voltage (typically 138 kV or 400 kV); higher voltage reduces the loss for the same transmitted power.

The voltage generated at the power source and the voltage that is suitable for the electrical load at the other end of the line will both be much lower than the line voltage. Voltage is stepped up at the generator end and stepped down at the other end of the line, via transformers. All machines that convert power from one form to another are less than perfectly efficient, as we saw in chapter 1, but it turns out that transformers are quite efficient machines, so that stepping the voltage up for transmission and then down for distribution is more economical than transmitting power without changing the voltage.

There is one serious problem with the use of transformers to step up voltage, however: they work only if the power is AC, not DC. I will not look into transformer operation here: there are many websites that discuss this important device (the operation is a simple application of induction). Nor will I say anything about how DC power is stepped up and down, or how DC is converted to AC and vice versa—we have too many fish to fry already. More directly relevant for us is *reactance* (another consequence of induction), which causes additional power to be lost when AC but not DC current is transmitted down a power line.

Current moving down a line constitutes a changing electric field, and so Faraday's law tells us that a magnetic field is induced, as we have seen. So far, so good. If the current is DC, the rate at which the field changes (i.e., the speed of the electrons moving down the line) is constant and so the induced magnetic field is constant. For an AC current, however, the electrons are jiggling back and forth, so the magnetic field they induce also jiggles back and forth. A changing magnetic field in turn induces an electric field. This induced electric field adds to the original electric field. (Complicated? Yes, and it gets more so, but don't sweat it—I will not be delving any deeper into electrodynamic technicalities.) The result is eddy currents, which reduce the current that moves in the middle of the conducting line and enhance the current on the outside of the line, as shown in figure 3.1. Thus, AC current is banished from the center of a conducting wire by induction. The effect is stronger at higher AC frequencies; at very high frequencies the electrons can travel only on the wire surface.

You can see from the figure that the cross section of wire through which the AC current can move is reduced compared with the cross section for DC current, and so the movement of electrons is impeded. The *impedance* of AC current traveling down a wire is the sum of resistance (a property of the conductor) and reactance (a consequence of induction). For a DC current, there is resistance but no reactance.

Power Transmission Engineering

With these electrodynamic preliminaries out of the way, we are now in a position to appreciate some of the technical issues that arise when large amounts of electrical power are transmitted along power lines. These issues are clearly important ones, given that most of a nation's power generating

FIG. 3.1. Resistance and reactance. (a) A DC current traveling down a conducting wire (*gray circle*) is here represented by electrons (*black dots*). They are distributed evenly through-out the wire. Their movement is restricted by 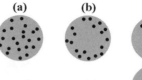 resistance of the conductor: the electrons bump into conductor material. (b) For an AC current, reactance effects restrict the electrons to the outer ring of the wire—a phenomenon known as the *skin effect*. As the AC frequency increases, the width of the ring of electrons decreases and the electrons be-come restricted to the wire surface. This restriction means that AC impedance can be many times greater than DC resistance, for the same power transmitted down the same wire. (c) Another phenomenon of AC currents, due to induction, is the *proximity effect*. When AC current moves along two wires that are close together, the electrons are re-stricted to areas as shown. The degree of restriction depends on how far apart the wires are as well as AC frequency.

capacity is delivered to the doorsteps of its citizens along electricity lines. Transmission networks are big business, as we will see, and the infrastruc-ture that constitutes them is everywhere; their maintenance is a significant expense and source of employment. Transmission networks are complex dy-namic systems, requiring constant monitoring, maintaining, and control-ling. The failure of any one of these can result in catastrophic loss of power.

The purpose of a transmission network or grid is to deliver electric power from a number of (usually remote) generating stations to a host of con-sumers. The power must be delivered with a constant voltage and current, or as nearly constant as can be achieved. With AC power, the system must march in step: the frequency must be constant and the phase must be syn-chronized across the entire network. In this section we will see why these difficult criteria must be met and how they are met—most of the time.

VOLTAGE

A transmission and distribution system is shown schematically in figure 3.2. We have seen that high-voltage power can be sent along a line with less heat loss (due to resistance in the conducting wires) than occurs with low-voltage power. We have also seen that it is much easier to step up and step down the voltage of AC electricity (via transformers) than DC electricity. These two facts of life mean that, with notable exceptions discussed later, most electri-cal power is transmitted as high-voltage AC. The generating station, be it a hydro dam up in the mountains or a nuclear station sited well away from

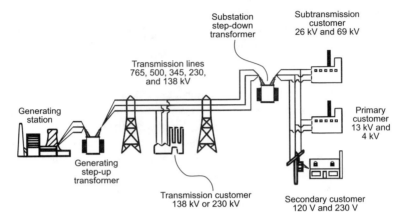

FIG. 3.2. A simple electricity grid. The generating station produces electric power at a potential of a few hundred or thousand volts, depending on the type of generator. For long-distance transmission to consumers, the voltage is stepped up by transformers in substations, to several hundred thousand volts. Near the consumers, the voltage is stepped down, in stages. Large consumers of electrical power, such as factories, tap into the grid at places where the voltage is high; domestic consumers are at the end of the line, where the voltage is maintained at either 120 V or 230 V (it varies from country to country). *U.S.–Canada Power System Outage Task Force (2004, p. 5).*

population centers, is a long distance from the consumers. Consequently, electricity needs to be moved hundreds of kilometers along power lines. The transformers that step up or step down the voltage along these transmission lines are huge—hundreds of tons—because the power in the lines is huge. Large transformers are very efficient (with efficiencies of at least 98%, and of 99.85% for superconducting transformers).[6]

There is an upper limit to the voltage that can be employed on utility lines. Increasing voltage makes it more likely that a *corona discharge* will occur. Air is normally a good insulator, and so, to minimize expense, high-voltage utility transmission lines are uninsulated.[7] (There are porcelain insulators to isolate the different lines, but the lines themselves are bare.) But at very high voltages the air can ionize, resulting in a discharge to a region of lower voltage, such as the ground. Such corona discharges are exactly like lightning bolts, and they can cause havoc throughout a network. (Discharge across an insulator is generally called *flashover*.)[8] In practice, the conducting wires of a transmission line are strung from tall towers so that the wires are far above the ground, thus reducing the likelihood of discharge.

Another reason for placing the wires high up is resistive heating. As current flows, it heats up a wire, which, being metal, then expands, becomes slacker, and droops toward the ground. Moreover, increasing the temperature of a wire further increases its electrical resistance. The power that can be stuffed down a transmission line of a given size thus depends on temperature. On a cold day, more current can be sent down a line than on a hot day. Modern technology allows utilities to monitor the temperature all along a transmission line and control the power accordingly. More on network control later in the chapter; this example serves as a hint that the transmission of electric power is a dynamic phenomenon that needs to be carefully watched —it is not a static system in which the utility company can simply flick on a switch and then leave things alone.

FREQUENCY

What frequency should be chosen for the AC signal? It cannot be too high because reactance losses increase with frequency. Also, electric motors run at speeds which require quite low frequencies. But the frequency cannot be too low because our incandescent and fluorescent lights would then flicker. These constraints leave quite a wide permissible range, and in fact the frequency that a nation adopts is largely historical. In the early days of electric power distribution, frequencies were much more varied than today, with different providers using their own preferred values independently of their competition. Over the years the nations of the world have settled upon one of two values for most of their utility power: 60 Hz (at 120 V) in North America and northern South America, and 50 Hz (at 230 V) almost everywhere else. A few countries, notably Japan, have both.

The nearly universal practice for transmission of electrical power is to provide three AC lines, each with the same frequency but with phases offset by a third of a cycle, as shown in figure 3.3. This *three-phase AC* design dates from the early days of large-scale electrical power transmission, having been introduced by Tesla in 1888. The three components travel in different wires along a transmission line, often with a fourth "neutral" wire (necessary for completing the electrical circuit, about which I will say no more here). A number of desirable properties of three-phase AC make it preferable to single-phase, or some other number of phase components:

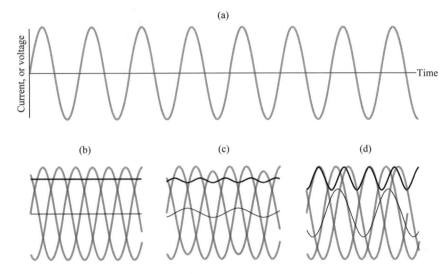

FIG. 3.3. Alternating current. (a) The magnitude of electric current at any given position along a conducting line oscillates in time at a specific frequency. (b) In three-phase AC three different currents are out of step with each other by one-third of a cycle (*gray lines*). If the magnitude of all three currents is the same, then the sum of the three currents is zero (*thin black line*), and the total power is constant (*thick black line*). (c) If one of the AC components has a different magnitude from the other two components (here, reduced by 10%), then neither the total current nor the total power are constant. (d) If the phase of one of the AC components is not right (here, off by 30°), then neither the total current nor the total power are constant.

- For a given voltage, three-phase AC requires less wire per mile than either single-phase or dual-phase.
- The transferred power is constant if the load is balanced (as in fig. 3.3b) and nearly so if the load is only slightly out of balance.
- Because the total current is zero if the loads balance (and small if they nearly do), the neutral wire can be rated for lower power and therefore be less expensive.
- Three-phase AC induces a magnetic field in one direction only (clockwise or counterclockwise), simplifying the design of electric motors.

At least three phases are required to produce all of the last three properties—constant power transfer, lower power for the neutral wire, and unidirectional magnetic field. Most household appliances require only a single phase, and in some parts of the world only a single phase enters a dwelling to power all the appliances in that household. We will see that it is important for a utility to *balance the load*, which is to say to draw power from all three phases equally as much as possible and match it to the power generated. That means that if a large number of houses are each fed a single phase, it should not be the same phase; instead, each phase should be distributed to the same number of houses.[9]

Grids

We saw in chapter 1 that over large areas the demand for power varies with the time of day and with location; one town switches on its street lighting at a different time than another, for example. The supply of power generated must match this varying demand. To do so, the utilities connect all their generators and power users (loads) in a network or grid that provides many routes from generator to load. This redundancy creates adaptability to changing circumstances. But while grids multiply the possible number of ways in which supply can be made to match demand, the task is a difficult one because networks also multiply the number of ways that things can go wrong. This problem of load matching is exacerbated by the very speed with which electrical power is transmitted and distributed over large areas. Power travels down a line at a speed that is within a few percent of the speed of light.[10] If a problem arises, it can spread over an entire grid (which may cover thousands of square miles) in a few seconds.

Power grids are not automatically stable, though a lot of thought and equipment has been dedicated to try to make them so. Power sloshes around a grid in different ways that depend upon the instantaneous load, so that the problem of balancing is a dynamic one. The fact that we use AC makes matters worse because we have to worry about *synchronization* of the power supplied by different sources. Let me explain.

Parts c and d of figure 3.3 give some indication of how power in a single line can vary when the phase is not right or when each phase does not have equal power. Variation in voltage and load on a larger scale—across a grid—can abruptly change the phases and the power transmitted along different lines, and these changes can cascade and amplify across the grid. The result is sometimes a massive *blackout*. For these reasons the power that is contributed to a grid from each generator must be synchronized with the power from other generators—the whole grid must beat with a common pulse. This requires precise timing and is easily upset.

Think what happens when, say, a short circuit occurs near a hydro generating station. Perhaps a truck crashes into a transmission tower, bringing down the lines and grounding them. There is a sudden loss of load—the machines that were drawing power further down the line can no longer do so—and yet power is still being supplied by falling water. The turbines at the generating station are still turning and dumping power onto the grid. The turbines suddenly find that there is no load holding them back, and so they speed up: it is exactly like releasing a brake. Because the turbines speed up, they generate AC at a higher frequency. Even if the fault is cleared (the short circuit is fixed), there is still a problem because now the generating station is out of phase as a result of its altered frequency.[11] This must be fixed or else the power oscillation problems hinted at in figure 3.3c and 3.3d will propagate across the grid.

A good analogy to help us think about the problems of maintaining a stable electrical grid system is to consider a large tank full of water with an open top. If there are no disturbances or leaks, then everything is calm; the water surface remains flat, and nothing is spilled out of the tank. If a disturbance occurs (say a large rock is dropped into the tank), waves move every which way and reflect off the sides of the tank, before eventually dissipating. If the tank is nearly full (corresponding to a grid operating near capacity), very little disturbance is acceptable because big waves will spill over the side and water is lost. Running a power grid is very much a balancing act, with quick responses to unforeseen changes being an essential part of the job.

BLACKOUT

A well-documented blackout, but by no means the world's largest, occurred during the afternoon of August 14, 2003, in several eastern and midwestern U.S. states and in the Canadian province of Ontario. An estimated 55 million people were affected, including the populations of New York City, Detroit, Baltimore, Cleveland, Newark, Buffalo, and many other cities. Houses, communities, and industries were out of power for two days (twice as long, for a small fraction of the total). An idea of the scale of the blackout can be seen from the satellite images shown in figure 3.4.

At its peak, more than 500 generators at 265 power stations were shut down; from 29 GW, the load dropped by 80%. Of course, domestic outlet power was lost, but so was water pressure, with the result that municipal water supplies became contaminated. Rail services were interrupted. Tele-

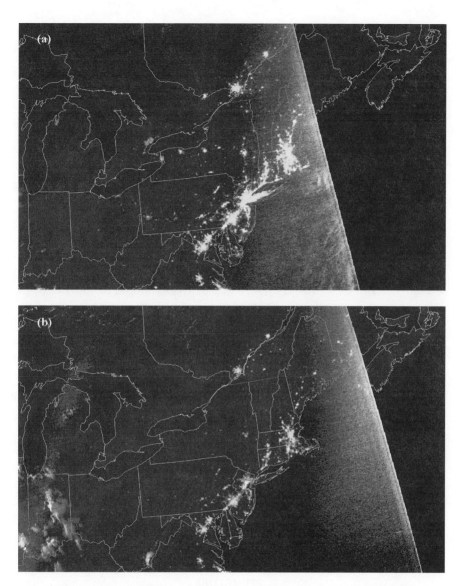

FIG. 3.4. The August 2003 blackout in the eastern United States and Canada. (a) Before (night of August 13–14). (b) During (night of August 14–15). *Images from the U.S. National Oceanic and Atmospheric Administration.*

phones continued to work, but the system became overloaded with calls reporting the blackout. Within the blackout region, there remained pockets of continued power where automatic control systems or backup generators worked. In most areas these failed, but some areas (such as Philadelphia) survived unscathed by rapidly disconnecting from the grid.

The instabilities of AC grid systems, alluded to in the last section, spread very rapidly because the speed of electricity is so fast: power generated in a plant at exactly 9:00:00 a.m. is consumed, perhaps hundreds of miles away, by 9:00:01 a.m.—within 1 second. Instabilities can travel as far, as fast. A joint U.S.–Canadian task force that investigated the cause of the 2003 blackout attributed it to a combination of circumstances and bad infrastructure.[12] While power was out, fingers were quickly pointed to the Canadian side of the grid, but the detailed report told a different story. High ambient temperatures led to heavy loads (from air conditioning) and sagging lines. Flashover from uninsulated lines to nearby trees resulted in short circuits; when these lines were bypassed, that increased the current on compensating lines. These overloaded lines overheated and tripped—and so the process cascaded across the grid. An Ohio-based grid system operator was found to have failed to maintain the system adequately and to have underestimated the potential (now that's a pun) for voltage instability. Available power sources in the Cleveland-Akron area were not brought on line to support the voltage when required. A software bug in the energy management system slowed down the system's response to initial instability.

This example of a blackout is one of many. The detailed cause or combination of causes may vary case to case; I mention this last (so far!) large North American blackout because it was so well-reported and investigated.[13] The main message to take from this event is the complexity of grid system dynamics and the sheer size of these grids. It is to grid structures that we now turn.

STRUCTURE

Each network or grid consists of many components that fall into broad categories: there are different types of power lines and cables, circuit breakers that differ in size and function, many types of switches, and millions of transformers of all sizes. From source to load a grid splits naturally into three segments: generation, transmission, and distribution. The network of lines from sources to loads is multiply redundant to enhance adaptability to changing loads and to unforeseen circumstances. Accurate timing permits synchronization across a very large grid. Three major grids cover most of the populated land surface of North America: the Eastern Interconnect (which suffered the 2003 blackout), the Western Interconnect, and the Texas Interconnect.[14] These grids are defined by their synchronization; thus, everywhere within the Western Interconnect (roughly, west of the Rockies) the AC phase is the same. Everywhere in the Eastern Interconnect the phase is also the same—but independent of the Western Interconnect.

It is desirable to share power between large grids to reduce expense (as we saw in chapter 1) and—quite literally—to spread the load. AC power cannot be transmitted across grid boundaries, however, because the grids are not synchronized. This problem is usually overcome with *high-voltage DC* (HVDC) lines, of which more below. Since DC has no phase, the conversion from AC to DC and back again is straightforward (via rectifiers), with no stability issues. Seven such HVDC lines link the Western and Eastern Interconnects, and two link the Eastern and Texas Interconnects. The world's largest grid, in terms of population served and power capacity, is the Continental Synchronous Area, which provides 400 million people in continental Europe with about 700 GW of electric power. It is linked to other, regional grids (such as those in Britain and Scandinavia) via HVDC.

Historically, large companies generated and distributed their own electricity. With the increase and spread of electrical power, vertically integrated utilities arose. That is, utility companies owned and operated all aspects of power generation, transmission, and distribution. Recently, this integrated structure has been changed (in the United States by law since 1996 and in practice since 2000). The idea was to free up different geographic regions to buy their electricity from the least expensive source. Long-distance transmission is cheap (typically about 2% of the cost to the consumer), so it is often cheaper to import power from some distant source than to grow your own. Nowadays there are many independent power suppliers, transmission companies, and distribution companies.

Deregulation was supposed to improve competitiveness and thus keep costs down. Unfortunately, it seems to have had the opposite effect and also has reduced system stability. The new companies did not invest in, or else insufficiently appreciated the need for, infrastructure such as *reactive power* generation. (Reactive power is AC power that has a phase difference of 90° between voltage and current. It does not contribute to useful work—that is, to power usage for which a consumer can be billed—but instead helps to maintain voltage across a grid.)[15] The new deregulated companies also did not report enough statistics to permit accurate forecasting or control.

New technology that will improve the situation with the next generation of networks, dubbed *smart grids*, is now on the horizon. I have more to say about smart grids in the last section of this chapter. Here, I will limit my comments to why they are needed. An example of automatic quick response to changing circumstances, so necessary for smooth operation of a power grid, is *automatic generation control*, which acts to maintain a constant AC frequency. We have seen that frequency is sensitive to the difference between source power being brought onto a grid and load power being taken off it; accurately monitoring the system frequency tells the grid operator whether he needs to increase supply or reduce it, and by how much. Despite this and other automatic controls, grid operators always have to battle instabilities which, as we have seen, can lead to catastrophic failures of power. Nations also worry about the vulnerability of power grids to sabotage.

With present technology, circuit breakers trip when the voltage or the current or the frequency doesn't look right. This approach kind of works, but it is clunky and inexact, to say the least. Utility operators today often resort to extreme, perhaps even desperate, measures to cope with system instabilities. One common example is *load-shedding*, or rolling blackout, in which local power is deliberately shut off as a last-resort attempt to forestall a full-blown blackout. A less-severe version is the *brownout*, in which system voltage is allowed to drop to prevent a blackout. Brownouts cause lights to dim and some equipment to fail. A sign of poor management or control, brownouts are very common in developing countries, and hardly unknown in even the most developed nations.

Couple these issues with an aging infrastructure (in the United States certainly—the present system was designed half a century ago—but also in many other developed countries), characterized by obsolete system layout and environmentally unfriendly components, and you can see why there is a push to next-generation smart grids. We will see that there is also a technological pull—the promise of inherently superior performance.

Transmission Losses and Limitations

We have seen the causes of power loss, via heat dissipation, along transmission and distribution lines. The numbers vary from country to country, averaging about 7% in OECD member states.[16] Thus, the combined transmission and distribution loss in the United States is 6.5%; it is 5% in Japan and 8.9% in Britain. (Interestingly, only 2% of the British loss is due to transmission line loss; the remainder is distribution loss.) Losses in less-developed nations are higher; India holds the unenviable title of world's least-efficient power provider, with a whopping 27% loss. Grid inefficiencies are the main culprit, but theft and corruption are important in some cases (in India "power pilfering"—overt tapping into overhead power lines—is rampant). How are these losses ascertained? Quite easily: the total energy produced in all generating stations at any given moment is known; this is compared with the energy sold at the same time. Any shortfall—whatever the cause—is deemed to be loss.[17]

The costs of such losses are readily imagined, given the huge amount of power that is carried on electricity grid systems. (For instance, with 15,000 power generators and a grid loss of 6.5%, the United States is wasting the output of nearly 1,000 plants.) Consequently, much effort has been put into reducing power losses. We have seen that DC is used to carry power between asynchronous grids; it is also used underwater, where it is more efficient than AC.[18] We have seen that the skin effect restricts the flow of AC to the periphery of a conducting wire; plaiting the wire in specific ways (*Litz wires, twisted pairs*) reduces the associated radiative losses. Other modified conductor geometries that mitigate loss effects include the common *coaxial cable*.[19]

The effects of losses are not confined to wasting electrical power; losses also limit the length of the lines that make up a grid. The length of a transmission line is limited by its electrical impedance; resistance effects show up in short lines and reactance effects in longer lines (recall that the electrical impedance of a conductor is the sum of its resistance and reactance). Double the length of a line and it loses twice as much heat for the same current flow. Let us say that a given line has low electrical resistance so that resistance is not the limiting factor. The length of this line is then limited by its reactance, which leads to an undesirable voltage drop. The magnitude of voltage drop equals current multiplied by impedance; this is why, even if electrical resistance in a line is low, the reactance can have adverse effects. Lines of intermediate length (say 100 miles) are limited by such voltage drop. Thus, losses limit the length achievable by individual lines within a grid. At the time of writing, however, the limit for grid size as a whole is not set directly by losses in the lines but instead by system instability effects.

The Underground Movement

One aspect of electrical power transmission affects everyone and yet is only tangential to our main subject. However, the *undergrounding* of power lines, as it is called, influences transmission losses, and so I think the subject is worth airing.

The motivation for placing power lines underground instead of stringing them between towers high up in the air is aesthetic as much as it is practical—aesthetic, because many people consider power lines, and their towers, to be an eyesore. Proponents of underground power lines often hitch environmental or health arguments to the issue as well. It may be that prolonged exposure to strong magnetic fields adversely affects human and animal health, but the magnetic fields surrounding transmission lines are not strong. The fields surrounding underground power lines are attenuated somewhat by the soil or concrete that encases the lines, but the argument for moving lines underground for this reason alone is weak, like the fields. A stronger case can be made for the adverse influence that overhead lines have on migrating bird populations: a small fraction of large migrating birds are killed each year when they collide with power lines. Aesthetically, most people would agree that overhead lines are very unsightly, especially when they proliferate in urban settings.

There are more pragmatic reasons for undergrounding that may influence hard-nosed utility companies more than do environmental or aesthetic concerns. Outages due to trees falling across overhead lines are very common, especially in winter in rural settings. In the Pacific Northwest, where I live, heavy autumnal rain softens the soil around tree roots, and then strong winds blow the trees over. We typically experience half a dozen power outages each fall/winter, and every time the power fails, my local utility company has to send out teams of engineers to remove the trees and fix the lines. These teams are on standby 24/7 throughout the winter months, such is the problem. Clearly, this is a major expense as well as a public relations embarrassment. Underground lines do not get cut by falling trees, nor do they get struck by lightning. Maintenance costs are lower than for overhead lines (no tree-trimming programs are required, for example). A European report suggests that, over the long term, both transmission losses and the price of electricity can be reduced by moving power lines underground; the predicted savings over the lifetime of an underground line may be two to four times the cost of installation.

So why are underground power lines so uncommon, except perhaps in big cities? Because the cost per mile of creating an underground line is always much greater than the cost of an overhead line between the same two points. Undergrounding costs vary widely (with geology and geography, for example), but they exceed the cost of overhead lines by a factor of between 2 and 10. For this reason alone, it is only in densely populated, developed nations such as Holland, Britain, and Germany that a significant fraction of the transmission lines are underground. Wealthy, technologically savvy North Americans have the know-how to put lines underground, of course, but they also have relatively large distances to cover, and that usually makes undergrounding uneconomical outside cities.

There are other reasons why overhead lines are preferred. The life expectancy of an underground line is half that of an overhead line; locating faults in an underground line is more difficult (so outages last longer, compared with overhead lines, though there are fewer of them); underground AC lines cannot be longer than about 30–50 km (longer lines must be DC); heat dissipation underground requires care; underground pipes require expensive insulation. For these reasons, ugly power lines and the towers that support them are likely to be with us for some time.[20]

HVDC

High-voltage direct current transmission lines are employed for special cases: we have already seen that DC's lack of phase sensitivity is an advantage when transferring power between asynchronous AC grids. DC is not used more widely because stepping up and stepping down DC voltage is expensive. On the other hand, DC losses along a transmission line are lower than AC losses because, you may recall, there is no reactance to DC flow, only resistance. Combine these two features—high conversion costs and lower losses—and you see that HVDC is viable for very long transmission lines. For such lines, the high cost of stepping up (at the start of the line) and stepping

FIG. 3.5. A transmission line on the Pacific DC Intertie, near the border of Nevada and California. Note the long insulators. *Photo from Wikipedia.*

down (at the finish) is offset by the saving in cable (which depends on line length) as well as power. In short, there is a break-even line length of about 400 miles.

Another instance where DC is preferred is for transmitting power to islands via undersea lines. The increased capacitance of these lines makes DC preferable to AC. In the future, transmission in the other direction—from offshore wind farms and wave power generators—will also be via DC. An early example of a HVDC line is the 43-mile 312-MW line that connects Vancouver Island, off the coast of British Columbia, with the mainland. A much bigger example is the Pacific DC Intertie, which sends electrical power from Washington State to Los Angeles. This intertie or interconnection, which dates from 1970, is an uninsulated overhead line (a pair of 40-mm diameter conducting wires) at 500 kV and 3.1 GW (3,100 MW). Thus, each wire carries a current of 3,100 amps. Given these figures, it is not hard to understand why the insulators on the transmission towers (fig. 3.5) are so long.

The biggest HVDC interconnection in the world stretches from a dam in rural China 1,250 miles to Shanghai; it carries 6.4 GW. Similar-sized HVDC systems are planned elsewhere in China and in India and Russia.[21]

The Future

We have no option but to continue using and further developing the electromagnetic power transmission infrastructure; there is simply no serious competition for bulk power transmission over long distances. Consequently, great effort is being made to improve this infrastructure via new technology. In this section we look at a few of the main ideas that are being implemented, or will soon be implemented when we learn how to do so efficiently.

SMART GRIDS

The electricity grid system of many developed nations is getting old and is not best configured for taking advantage of digital technology and new sources of renewable power. I have described how utilities can respond to changing demand, but the response is slow and clumsy compared with what modern intelligent electronic systems are capable of providing. Smart grids will fine-tune response to changing demand and optimize the power provided to individual consumers (i.e., minimize supplied power without compromising demand).

At present some 90 smart grid pilot projects are in operation around the world. These aim to educate both supplier and consumer. The power industry considers selling the idea of smart grids to the public as vital to success. To this end, there are many books, magazine articles, TV ads, and video clips you can access in order to learn about the subject.[22] Why is the hard sell considered necessary? The front end of smart grids is the *smart meter*, which, even as you read this book, is being installed at your house or a neighbor's. These meters have much more capability than the clunky old meters. A smart meter will monitor your power use hourly or more frequently and download this information to your electricity provider. It is this very capability that raises the concern of some civil liberties groups.[23] Some people see this as Big Brother peeking in through your window, monitoring your life and changing it as he wishes. Smart meter communication is two-way, so that in principle your public utility could switch off your washing machine during a period of peak power demand and switch it back on again during the night, when rates and demand are cheaper. In time, householders will learn how best to utilize their electricity and so reduce their costs and those of the utility. Clearly, a process of education, for both sides, is necessary to ensure that such power management fine-tuning does not step over a privacy line.

Another capability of smart grids, also linked with improved matching of supply with demand, is their ability to connect the many small and widely dispersed sources of renewable energy that are anticipated to come online within the next few decades. Thus, the expected large number of hybrid cars and electric vehicles will be able to contribute to the grid when plugged in for recharging. Millions of batteries will provide electricity to the grid (for a price) during periods of high demand and will charge up (for a lower price) at night or during the day at times of low demand—this is *V2G carbitrage*, you may recall. A sophisticated and flexible management system will be required to control such a two-way flow of power and money.

Wind and solar will also contribute power to the smart grid. Wind farms are often located far from population centers, and they produce power at a variable rate because of the vagaries of the wind. Thus, integrating a wind farm's electrical power supply into a grid requires just the kind of fast and flexible management that smart grids can provide. Solar panels on rooftops will similarly feed into grid supply. These and other renewable fuel sources can enable smart grids to reduce a nation's carbon footprint (power generation currently contributes 40% to the U.S. carbon footprint).

Smart grids will better ensure grid stability by faster and more intelligent responses to accidents, acts of terrorism, or freaks of nature. Thus, they will, when fully implemented, provide all-round better service: reduced demand for the same supply, greener power, and improved stability and reliability (or "self-healing," as it has been called). Analysts point out that it is not necessary for all the proposed improvements that lead to a smart grid to be implemented at the same time: the system will improve gradually as one or another factor is brought online. Indeed, smart grid applications have been gradually increasing since the 1980s, in industrial plants, for example. The big change, however, will be seen when the new technology is installed in our homes.

All these improvements—smart meters, more capable system sensors, more and reconfigured grid infrastructure, and so on—come at a considerable cost, of course; hence the ads. The National Energy Technology Laboratory (part of the U.S. Department of Energy) estimates that the savings brought about by smart grids will amount to between four and eight times the initial cost (presumably over the lifetime of the new grid infrastructure).

SUPERCONDUCTING TRANSMISSION LINES

Superconducting materials have an amazing and very counterintuitive property: electrons (and hence electrical current) can flow through them with zero resistance. Not a very small resistance, please note, but *zero*. Zip, zilch, nothing at all. A weird phenomenon that was restricted to the physics lab, superconductivity became technologically interesting in the 1990s with the discovery and development of so-called high-temperature superconductors. The problem with early superconductors was that they worked only at temperatures that are within a degree or two of absolute zero ($-273°C$, $-459°F$). It is extremely difficult and expensive to generate such low temperatures, so superconductors, while interesting, found no application in any industry with a bottom line. Some of the new copper oxide high-temperature superconductors work at temperatures as high as $-140°C$. While still very cold, this temperature is inexpensive to generate because it is above $-196°C$, the temperature at which liquid nitrogen boils—and liquid nitrogen is relatively cheap to make. In other words, some high-temperature superconductors can be created by cooling a copper oxide material with liquid nitrogen. This makes underground superconducting lines just about commercially viable.[24]

The advantage of zero resistance to current flow is obvious: normal electrical resistance increases with conductor length, but if the resistance is zero, then it doesn't matter how long the conductor is. So, DC current can flow along a superconducting line of any length without resistive loss. (AC current will still suffer from reactance loss, as we have seen.) Consequently, superconducting transmission lines could be useful for carrying electrical power from remote locations (such as out-of-the-way wind farms) to the grid. A few superconducting transmission lines are in service today, including a 574-MW line on Long Island, New York. Even though the liquid nitrogen cooling makes the transmission line itself more expensive than conventional lines, it is actually more economical for transmitting power because the superconducting wire can carry 150 times the power of a conventional conductor of the same diameter. One advantage that is seen as very significant for the future, when superconducting lines are envisaged as forming a significant fraction of transmission grid systems, is the self-regulating nature of superconductors when subjected to power surges. When the flow of current becomes excessive, a superconducting line will lose its superconducting capability; it becomes normally resistive and will resist flow, thus correcting the surge. So, a superconducting grid will be more stable than are existing transmission grids.

WIRELESS TRANSMISSION

Wireless transmission sounds futuristic, and it is, though wireless power transmission has been with us for some time, albeit over short distances. Inductance coupling, for example, transfers power between a rotating flywheel in a turbine and a conducting cable via induced magnetic and electric fields. Such wireless transmission of power covers only a few inches or feet, however; to transmit power over many hundreds of miles calls for an altogether different technology.

In 1975, NASA scientists demonstrated that power could be transmitted through the air from generator to *rectenna* receiver. (A rectenna is an antenna that *rectifies* the received signal—turns it into DC.) At NASA's Goldstone facility in the Mojave Desert, microwaves were transmitted to a receiver 1 mile distant. The power of these microwaves was 34 kW, and the efficiency of the transfer was determined to be 82%. This demonstration was at the time, and remains today, the longest distance that such power has been transmitted wirelessly (so we are told; I suspect that classified Star Wars research could tell us a different story).

The idea that electromagnetic radiation can be used to transfer power from *A* to *B* is hardly a surprise: the sun has been doing it for eons, and NASA has been giving the concept serious consideration for at least 40 years. The technological challenge is to make such transfers efficient and commercially viable. One very futuristic scheme that promises clean and plentiful energy is to place large solar panel arrays in space and beam down microwave power to even larger rectenna receivers on the earth's surface. I summarize and investigate this idea in chapter 8.

The laws of electrodynamics and the physical phenomena these describe, such as resistance and inductance, shape the characteristics of power transmission. Losses depend upon voltage, line length, line structure, and transmitted frequency, with DC losses being lower than AC losses (but DC is more expensive to step up and down). Power grids are huge, expensive, complex, and difficult to keep stable. Instability results from the speed of electricity and from the characteristics of AC transmission (in particular, the need to maintain phase synchronization across the grid). Brownouts and blackouts are consequences of grid instability. Deregulation has complicated the stability problem. New technology such as superconducting lines and smart grids will improve grid performance in the next few years. At the same time, such improvements are more urgently needed because of the increasing number of small, widely spread renewable energy sources such as wind farms and electric vehicles, which have variable power output.

OLD KING COAL

Coal has a black image. Just as some people hate cigarettes but depend on them, so humans for centuries have had a love-hate relationship with coal. Coal mining and coal burning are dirty, unhealthful, ugly, bad for the environment, and vital for industry. Coal kills people at every stage of their involvement with it, from extraction to combustion, yet it has kept them warm in winter when no other fuel was available. As with tobacco, coal has powerful self-interested advocates who remind us just how many jobs the industry provides and how much revenue it generates; at the same time, its powerful and vociferous enemies show us graphic images of clogged lungs and death.

Sorting out the hard facts from the emotion is as necessary as it is difficult. In this chapter I look at the place that coal occupies in the world of power generation, today and in the future.

The Role of Coal

Given the increasing world population (our numbers reached seven billion during the writing of this book) and the rapid industrial development of populous countries such as China and India, it is reasonable to assume that our voracious appetite for energy will increase during the coming decades. Global consumption of coal increased by one billion tons between 2001 and 2005; across the world, coal-fired power stations are opening at the rate of one per week. Worldwide, 40% of electrical power is provided by coal; it is closer to 50% in the United States, 70% in India, and 80% in China. Long gone are the days when coal was a declining industry (due to oil). Its nadir was back in 1961; in the 47 years between then and 2008, the production of coal increased 255%.

It is sensible to ask how much fossil fuel remains in the ground for us to exploit during the foreseeable future. For environmental reasons we may not want to burn such fuels, but until we develop alternatives, we may have no choice. So what are the numbers? A telling figure is the ratio of *proven reserves*—known underground fuel resources that can be extracted economically—to current consumption. Despite its inexactness, the ratio for 2010 provides a plausible estimate of the number of years a given fuel resource will last. The ratio for oil is 49; for natural gas, 59; and for coal, 161. Thus, we will run out of oil and natural gas before the end of the century, and unless we find a better (larger, more economical, and cleaner) fuel by then, we will have to fall back once again on coal as our principle source of energy. Both of the figures that produce this ratio may change in the future: reserves will increase with further prospecting, and proven reserves will increase with technological advances; consumption will increase with increasing population and industrialization.[1]

In figure 2.6 we saw that coal provides much more of our electricity than any other fuel source. The figure today is 39%, and 30-year projections do not see this figure changing much. (Natural gas is currently in second place, at 19%, and will increase to around 30%.) In addition, coking coal is used for making two-thirds of the world's steel. Another indicator of the significance of coal, one that sheds light on the economics of coal extraction, comes from western Canada. This area is a leading producer of natural gas (and of its sibling, oil), and yet even here, the cost of energy derived from natural gas is $6 to $14 per gigajoule (the price is volatile), whereas the same energy provided by coal costs $1 (and the price is relatively stable).[2]

All these hard facts tell us that the coal industry is in the leading position for providing us with energy for the next several generations unless new technology provides us with a clean and economic alternative. Even if an overwhelmingly superior new technology were to appear tomorrow, coal will remain with us for a long time because of its low cost and abundance and because of the deep penetration of the industry into the economic infrastructure of many countries. For these countries, getting rid of coal would not be like discarding a hat; it would be more like having a leg amputated. The role of coal is crucial.

Another consideration, only partly economic, is energy security. Coal is seen as a safe source of energy in that it is reliable (there are many friendly nations that export coal). The reasons are straightforward enough to simply list:

- Coal supplies do not require pipelines or dedicated supply routes that are expensive to protect.
- Coal is easily stockpiled in times of need.[3]
- The United States was self-sufficient in energy until 1958; since then, because of our growing expenditure of energy, the gap between domestic supply and demand has widened more and more, as is illustrated in figure 2.8. Available domestic fossil fuel sources are almost entirely coal, at 94% (such is the whim of geology).
- The United States is second only to China in the production and consumption of coal.

COAL CALCULATIONS

One kilogram of coal burns to yield 6.67 kWh of heat energy. This energy density is 150% that of firewood, which helps explain the popularity of coal in days gone by. (The energy density of oil is 11.5 kWh/kg, but oil is much more expensive than coal.) The carbon content of coal varies from about 75% for lignite to 95% for anthracite; let us fix an average value of 85%.* How much carbon dioxide is generated for each kilogram of coal burned? Carbon has an atomic weight of 12 and CO_2 has an atomic weight of 44, so 1 kg of coal burns to produce 3.1 kg of CO_2.

If your house is powered by electricity from a coal-fired power station, how much CO_2 do you produce when cooking your Thanksgiving turkey? Say you cook the bird for 4 hours; a typical oven draws 1.35 kW power, so cooking consumes about 5.2 kWh of energy. A coal-fired power plant is typically 35% efficient. Therefore, to cook your bird, you must burn about 2.2 kg of coal, and thus generate 6.9 kg of CO_2. That's 15 pounds, or about the same weight as the turkey. (This is likely to be a low estimate because I have neglected transmission line losses.)

A popular calculation of this type, designed to convey coal consumption for everyday activity, is to determine the amount of coal needed to keep a 100-W light bulb lit for a year. The bulb consumes 876 kWh of energy during this time, and so requires 375 kg of coal (875 lb). Burning this amount of coal generates 1,163 kg of CO_2. That's more than a ton. At standard temperature and pressure, CO_2 is twice as dense as air, and so our light bulb generates about 587 m^3 of carbon dioxide (20,000 ft^3). If your power comes from a coal-fired plant, and you want to reduce your carbon footprint, consider turning off the lights when you leave a room.

*The figures for the carbon content of coal are from Harvey (1986).

For all these reasons, coal industry promoters in the United States cast coal (now that's a pun) as the patriotic fuel. Needless to say, the predominance of coal at home means that there are a lot of jobs in the coal industry. The combination of energy security, jobs, and costs provides potent arguments for politicians and business interests, if not environmentalists, for protecting the coal sector.

Pollution, Health, and Safety

POLLUTION STATISTICS

Coal has never been pretty. Extracting the stuff from Mother Earth leaves scars over the land (as well as on miners—I will get to the perils of mining soon enough). Burning it blackens our cities. In countries that industrialized early, such as Britain, soot-encrusted, smoke-cloaked grimy cities became an icon of the evils of the Industrial Revolution, well described by writers of the day who had a social conscience and a social agenda, like Charles Dickens. The same smoke and smog lingers over cities today in industrializing countries such as China, India, and Brazil. Television coverage of the Beijing Summer Olympics in 2008 made the Western world well aware of Beijing smog.

London fog is infamous, thanks to Sherlock Holmes as well as the "killer fog" of 1952. The well-known bad British weather was only partly to blame for the 1952 phenomenon: nineteenth- and early-twentieth-century fog was in fact smog, generated by smoke coming from millions of domestic coal fires. On December 5, 1952, unusual weather conditions trapped the smog; and for months afterwards, the mortality rates in London were well above normal. At the time, this increase was ascribed to the flu, but subsequent studies showed that 3,000 deaths were attributable to the air pollution. A more recent retrospective study suggests that the true number of deaths might have been several times greater than this.[4]

An environmental study published in 2000, sponsored by the Clean Air Task Force (a nonprofit environmental group), concluded that in the United States 30,000 deaths per year are caused by air pollution. This number is about the same as the annual number of deaths due to firearms. A state-by-state analysis of the data showed that those states with coal-fired power stations fared the worst, thus demonstrating, according to the study, that the particulate matter from coal burning is deadlier than that from other sources, such as car exhausts. The report's authors estimated that 60% of the deaths could be prevented by reducing emissions of sulfur dioxide and nitrogen oxides from power plants. And apart from the estimated 30,000 deaths, there are many times more illnesses and exacerbated respiratory conditions reported.[5]

Particulate matter in the air clogs lungs and kills people: of that there is no doubt. The problem with the two statistical analyses I have summarized here is that they are just that—statistical. Most people have a poor intuition about statistics, and because statistical studies produce correlations and not direct cause-and-effect evidence, we tend to dismiss them. An industrial disaster that killed 30,000 Americans in one fell swoop would be noticed at once, and action would be taken to ensure it did not happen again, but an invisible killer that acts more slowly over a wider area is barely perceived. So here is an observation that may more easily get the attention of many people (I am once more beating my drum about perceived risks): coal ash is about 100 times more radioactive than nuclear power plants. That is to say, the usual emissions from a coal-fired power plant are much more radioactive than the usual emissions from a nuclear plant of the same capacity (the qualifier "usual" is significant). Coal-fired power plants do not generate radioactive waste, but they do release into the air radioactivity that is naturally present in the coal. Uranium and thorium, two radioactive elements, exist in many soils, rocks, and minerals, including coal. The U.S. Geological Survey points out that the level of radioactivity in coal-burning emissions is no cause for alarm, but I would be willing to bet that if it caused 30,000 deaths per year, it would generate a lot more alarm than coal-induced smog.[6]

MINING MORTALITY

Apart from air pollution, coal leads to many deaths each year from just extracting the stuff out of the ground. Coal mines miners—less so today than a hundred years ago but still at the rate of thousands of deaths per year around the world. Coal mines kill with poisonous gas (such as hydrogen sulfide), explosive natural gas (methane, known to miners as *firedamp*), mine tunnel collapse, and flooding. In the five-year period from 2006 to 2011 there were 177 fatalities in U.S. mines, of which two-thirds were underground in deep mines.[7] The year 2010 brought three very-much-publicized mining disasters. In April, 29 West Virginia men died at the Upper Big Branch Mine, despite several recent evacuations due to concerns about methane levels. In August, 33 Chilean miners were trapped underground for months following a cave-in at the old San José Mine, all eventually being rescued via a specially dug escape tunnel. In November, 29 New Zealand miners died in an underground explosion at the Pike River Mine. Perhaps it is significant that the San José is a copper-gold mine whereas Upper Big Branch and Pike River are coal mines.

Today, the worst fatalities among coal miners occur in China. In part, this is because there are a lot of coal mines in China, but it is also due to unregulated mining, corruption, or low safety standards in regulated mines. Officially, the Chinese coal mine death rate since 2000 has been 4,700 miners per year, but this number is probably a drop in the coal bucket; it is certainly an underestimate of the true figure because small local mine operators usually do not publish details of accidents for fear of being shut down or, in the case of illegal mines, of being discovered and prosecuted. Accepting the official figures, China produces one-third of the world's coal but four-fifths of the annual mining fatalities.[8]

GREENHOUSE GAS

Certain gases in our atmosphere trap the sun's radiation and lead to a rise in temperatures around the world. The main gases responsible for this greenhouse effect—which leads to global warming—are water vapor and carbon dioxide.[9] So far, so good; a warm atmosphere has helped life to proliferate on earth. Too much of a good thing, however, is bad.

An increasing number of experts believe that man-made greenhouse gases (principally carbon dioxide) are leading to an overheating of our planetary atmosphere, with possible dire consequences further down the line if the overheating continues unchecked. There are vested interests in government and industry who would like all this to be not true and who have vociferously rejected the idea that mankind is responsible for the undisputed recent rise in atmospheric greenhouse gases; they say that these rises are part of a natural cycle. I used to be skeptical about doomsaying environmentalist claims regarding the evils of industrialization: ozone holes, global warming, and the like. (After all, it was only a generation ago that some of the same people were warning us about "nuclear winter"—an apocalyptic *cooling* of the planet due to atmospheric testing of nuclear bombs.) The evidence now accumulated, however, is compelling and suggests that there is an abnormally high concentration of greenhouse gases in the atmosphere and that this rise is due largely to industrial output. The main greenhouse gas is carbon dioxide; the main industrial contributor to CO_2 increases is coal-fired power plants.

First, let's look at the evidence for increasing atmospheric CO_2. There are a number of ways of monitoring gas concentrations in the air, and modern measuring techniques are very accurate. It is also possible to estimate with small error the historical concentrations of greenhouse gases by analyzing the constituents of air bubbles found in ice core samples taken from the polar ice caps. Crucially, these cores can be dated—they stratify like rocks—and so atmospheric gas concentrations going back centuries can be reconstructed. Also, the total can be broken down into a natural contribution arising from biological or physical processes occurring on the surface of the earth, and man-made contributions arising from burning fossil fuels that originate beneath the surface. (The two contributions—surface and subsurface—have different proportions of carbon isotopes; these can be measured accurately.) The result is shown in figure 4.1. Carbon dioxide concentration in the air has increased some 30% since preindustrial times and is rising.[10]

To place the anthropogenic contribution in context, we should consider the natural sources of atmospheric CO_2. Animal and plant respiration, the decay of dead organisms, and volcanic activity all generate massive amounts of CO_2. Indeed, compared with these, the man-made contribution is a mere 3%. In addition to the *sources* of CO_2 we should also consider the *sinks*: carbon dioxide is removed from the air by photosynthesizing plants and by water. The carbon that makes up the bulk of plant matter comes from atmospheric CO_2; it is taken in through the surface of countless leaves from countless trees during photosynthesis.[11] Atmospheric CO_2 is dissolved in seawater and lake water, thus removing it from the air. These processes—respiration and photosynthesis, biological decay and absorption by water, plus sedimentation—link together to form the *carbon cycle*, a gigantic and complicated series of processes in which carbon travels around large loops, like a giant Ferris wheel slowly turning. Atmospheric CO_2 is absorbed by plants to form trees, which rot and give up their carbon to the air. Or trees that fall down get covered with sediment and eventually form coal beds, fixing the carbon in the ground—until the seam gets exposed and weathering releases it again. Or the oceans absorb CO_2, which is then ingested by oceanic life, which dies and sediments out, and so on.[12]

I mention the carbon cycle because it is a natural cycle. Carbon is moved around the oceans, crust, and atmosphere of the earth via a series of loops with different timescales and in different degrees. There is a natural balance, a default level of carbon dioxide in the air, resulting from the carbon cycle. There will be annual wobbles in the airborne concentration of CO_2, as we

(a)

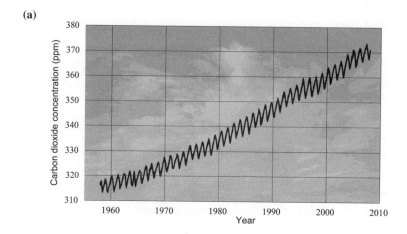

(b)

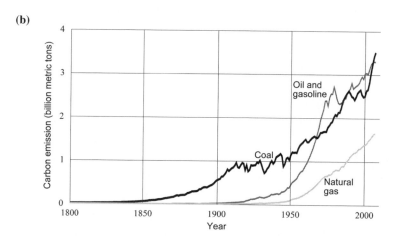

FIG. 4.1. Atmospheric carbon dioxide. (a) Direct measurement of CO_2 concentration in the air (parts per million by volume), made for over half a century at Mauna Loa in Hawaii. Annual variations are apparent; the rising trend is also clear. (b) The contribution to carbon emissions from burning fossil fuels. The rise began with the Industrial Revolution in Europe. Today, coal burning just barely surpasses oil and gas burning as the largest contributor. *(a) Data from the National Oceanic and Atmospheric Administration; (b) data from the Carbon Dioxide Information Analysis Center, Oak Ridge National Laboratory.*

saw in figure 4.1, and other oscillations leading to a fluctuating graph of CO_2 concentration when plotted against time. However, the human contribution may throw the system out of whack, upsetting the balance by generating in an instant (a couple of hundred years is an instant to a geologist and an evolutionary biologist) a huge surplus of atmospheric CO_2. Even though mankind's contribution is only 3%, it is 3% that is not obviously balanced by the available natural sinks.

These sinks will probably step up to the plate, in the fullness of time, and absorb the extra CO_2, taking the excess in stride so that the carbon cycle is maintained. But because the process of cycling carbon is so complicated, the sudden dumping of extra CO_2 into the system may cause large, unpredictable disturbances with the potential for very bad effects, such as changing ocean currents and weather patterns, or drastic temperature alterations in different parts of the world. Think of a massive spinning top that is stable— barely wobbling—until it is given a sharp shove: the consequence may be wild and violent oscillations. The top may eventually return to equilibrium as the oscillations die down, or it may fall over. The world will adjust to the shove we have given it—the sudden dumping of carbon dioxide—but it may become less habitable in the process.

The debate about global warming and carbon footprints has massive political and economic ramifications, of course, and leads to a lot of verbal as well as physical hot air. It seems sensible to reduce the magnitude of CO_2 that we dump into the atmosphere, given both our measurements and our understanding (and, perhaps more importantly, our lack of understanding) of the possible global consequences. This reduction can be achieved by reducing CO_2 emissions worldwide, or by increasing CO_2 sinks, or by doing both, and we will see how these can be achieved in a later section. Here, I will simply note that it is proving difficult, to say the least, to obtain a consensus and to generate action. Some people choose to deny that there is a problem. Some people in developing countries get annoyed when those in developed countries demand reductions in carbon emissions. ("You spewed out CO_2 during your industrial revolution, and now that you see its bad effects, you ask us to foot the bill by not having an industrial revolution of our own!") Some industries in developed nations consider that they are being unfairly targeted, and many point out the adverse impact on jobs or electricity costs or GDP of imposing emission reduction quotas, and so forth.

That the present state of carbon emissions is unsustainable in the long term becomes crystal clear with another quick calculation, with which I conclude this section. You will note that the calculation points out the problem but in no way suggests a solution. Finding the solution is a key aspect of solving the world's energy problems.

There are some 980 billion tons of coal reserves known to exist around the world. These will be used up within the next blink of an eye (i.e., within the next century or so, as we saw in the first section of this chapter). Using the methods adopted earlier in the sidebar "Coal Calculations," we can readily determine how much CO_2 these reserves will add to the atmosphere unless steps are taken to reduce emissions: about 2,200 billion tons. The current level of atmospheric CO_2 is 3,160 billion tons, so coal alone (we are ignoring oil and natural gas) will increase the amount of carbon dioxide by 70% unless it is very rapidly absorbed by the oceans. Let us guess that the oceans are capable of sopping up 99% of this excess atmospheric carbon dioxide, and quickly. Given that people are worrying about current increases of 50 ppm (parts per million; see fig. 4.1a), what will an increase of 7,000 ppm do to our planet?[13]

The Smil Drill

To find something positive to say about coal as an energy source, apart from the obvious and compelling economic argument, I must turn to what I call the "Smil drill." Vaclav Smil, a well-respected expert on energy infrastructures, has emphasized the concept of *power area density* as an important factor to be considered in the energy debate.[14] This is a revealing statistic when calculated with care and permits comparison across power generation technologies. Coal is something of an embarrassment for some of its supporters because it is so obviously dirty and unhealthful that proponents are seen as being motivated by economic gain alone: apart from financial self-interest, why else would they support such a polluting source of power? Well, one reason is power area density.

Power area density is the power produced at a production site per unit area. Evaluating it for a given industry is more complicated than you might think. For example, the mine from which the coal is dug occupies a certain area of land (more for a surface or strip mine than for a deep mine). In addition, there is the energy density of the coal to consider, the distance it must be transported to the coal-burning power plant, the power plant's efficiency, the area taken up by coal storage yards and the area of the fly ash disposal sites. It gets more technical: coal power generation requires flue gas desulfurization and on-site ponds for storing the resulting slurry.

Smil ends up with a range of figures that depend upon the details of a particular mine and power plant. He concludes that for coal, the power area density figure varies between about 100 W/m² and 1,000 W/m². That is, for each square meter of land taken up by the industry (the whole of the industry for a given region, from extraction to burning and by-product disposal), the power generated, net of production power requirements and the power required to dispose of waste, is several hundred watts. You can see that the calculation requires a detailed knowledge of the industry from the bottom of the deepest mine to the top of the highest smokestack.

The point of the exercise is that we can compare power area densities across technologies. In table 4.1 you can see Smil's results for coal and other biomass power plants and for renewable energy industries such as solar and wind power generation. (In chapter 7 I will apply the Smil drill to the nuclear power industry.) In each case the calculations require knowledge of the technical details and of how power is generated for a given technology.[15] The results are rather depressing, as they show one of the costs that must be incurred if a nation chooses to replace coal-powered generators with greener, renewable sources of power such as wood, wind, and sunlight. Because fossil fuels such as coal and natural gas take up much less space than renewable technologies generating the same power, they have much higher power densities than the renewables. The comparatively high power area density of coal means that turning to renewables would require a nation to dedicate 10–1,000 times the land area to power generation. In some countries this land cost can be absorbed, but in densely populated developed countries such as those in Europe, the idea of giving over so much extra acreage to the power industry is unrealistic.[16]

One renewable technology that is notably absent from table 4.1 is hydroelectricity. This absence is not really an oversight, despite the undoubted

TABLE 4.1. Power area density for different power generation industries

Power source	Power density (W/m²)
Natural gas	200–2,000
Coal	100–1,000
Solar, CSP[a]	4–10
Solar, photovoltaic	4–9
Wind	0.5–1.5
Wood	0.5–0.6

Source: Adapted from Smil (2010b).
[a] CSP = concentrating solar power

importance of hydropower in some regions and its deserved reputation for relative cleanness, because it is so site-dependent. Geology limits dam placement, but so does population density: dams cannot be built in densely populated regions without generating massive social disruption along with gigawatts of power, as we will see in chapter 6. In Western countries at least, hydropower can be significant only in territory with little population. Consequently, the power area density of hydropower is not going to be a determining factor. For the record, it is likely that power area densities for the hydropower are pretty low, given the large area of reservoirs that build up behind dams.

Thus, a reality of coal power generation (for some people, an uncomfortable one) is that it takes up much less land area than other power technologies.[17] I can imagine a coal industry spokesman jumping on this fact, and even selling his industry as green and environmentally friendly by virtue of it (telling us with a straight face that "more land can be used for planting trees"). It is hard to defend coal as an environmentally responsible source of fuel, but power area density is a factor that must be considered in a world of increasing population.

Mining

The economic feasibility of extracting coal from a given site depends on many factors—the depth and grade (quality) of the coal, the quantity of coal and seam thickness, the local geology, surface topography and groundwater conditions. Coal seam depth influences the method of mining: surface or underground. If the coal is located at a depth shallower than 100 m, surface mining (also known as *open cast* or *strip* mining) is usually adopted; otherwise, deep underground mining is the preferred method. There are exceptions: in Canada most coal is over 300 m below ground, and yet more than 90% of it is mined in open cast pits. Surface mining takes up a lot more ground area than underground mining, but in Canada space is not usually a problem. Globally, 40% of mines are open cast (67% in the United States). Mining from the surface extracts coal more efficiently and less expensively, as we will see, but deeper coal is usually higher grade, and sometimes underground mining is unavoidable.[18]

Surface mining is done in a series of strips, side by side. The *overburden* (dirt and rock lying above the coal seam) from the first strip is removed and trucked to an external site for disposal. The exposed seam is drilled and fractured by explosives, and mined. When the first strip is exhausted, miners move to an adjacent strip. The overburden is removed and dumped into the exhausted strip, and the process is repeated as before. The machines that remove the overburden and mine the coal seam are often enormous, be they dragline excavators, bucket-wheel excavators, or power shovels.[19] The world's largest bucket-wheel shovel is Bagger 293, which operates in a lignite mine near Hambach, Germany (fig. 4.2).

When it has been removed from the ground, coal is processed before being shipped to customers. In a typical modern coal-processing plant, the coal is removed from the site by conveyor belts to the plant, where it is separated from rocks and dirt in a flotation tank. (Coal is less dense than the unwanted material and floats to the top, where it is skimmed.) The wet coal is dried in a giant spin dryer, then crushed and dropped into a hopper, which fills railcars or trucks. It is then transported off-site. The whole processing stage, from conveyor belts to railcars, takes only 15 minutes. The

FIG. 4.2. Bagger 293, the world's largest bucket-wheel shovel. This gargantuan machine is 740 feet long and weighs 14,000 tons. The wheel has a diameter of 66 feet, and the buckets can dig up 8.5 million cubic feet of overburden or coal per day. Bagger 293 is operated by a crew of five. *Image from Elsdorf-blog.de.*

water that was used to separate the coal contains contaminants and is disposed of (carefully, we hope) as slurry.[20]

In recent decades, automation has made coal production much more efficient. Whereas 30 years ago, 230,000 American miners dug 800 million tons of coal each year, today 100,000 miners dig 1,100 million tons. Surface mining is more efficient than underground mining because almost all of the coal in a seam can be removed. This is not the case underground.

Let me explain. For underground mining, the overburden is left undisturbed, and the coal seam is mined beneath it. Some coal pillars must be left in place to prevent the roof from caving in. The fraction of coal that must remain behind as a roof support varies with the type of coal and local geology, but it is typically 40%. I have just described the *room-and-pillar* method of underground mining. This method works for coal seams that are horizontal, or nearly so. Another method is *longwall mining*. Here, a giant machine slices a 1-m or 2-m layer of coal off a long stretch of exposed coal face, and the coal is dumped directly onto a conveyor belt that transports it away from the face and up to the surface for processing.

In underground mining, hydraulically operated shields and roadway supports protect miners from roof falls. Holes drilled in the coal face connect to pipes which suck away firedamp. Water is sprayed at the coal face to reduce the risk of coal dust explosion and to protect the miners from pneumoconiosis, or black lung disease (formerly a scourge of mining communities but now almost eradicated). All these precautions add cost to underground mining, cost that is mostly absent from surface mining.

For underground mines especially, mine operators have to think carefully and in three dimensions about the best way to safely and economically extract the most coal from a seam: where and how long the tunnels and seam face should be; when to stop digging and move on, exposing a new face of the coal seam; when and how to collapse an exhausted section. The understanding of mining, computerized automation, and the huge machines have all led to improvements in productivity, but this will not continue forever. *Peak coal*—the year when world coal output reaches a maximum—is estimated to be only two or three decades away. Thereafter, miners will be pursuing an increasingly scarce, hard-to-get-at resource that will dwindle and become uneconomical to dig out of the ground. By then, let's hope, we will have some other energy source that may begin to replace it.

Burning

If only we could set aside the polluting nature of coal burning, coal would be seen (and is seen, by some people) as a very attractive fuel for electrical power generation. I have noted several times that it is inexpensive and reliably sourced—that is, it is locally abundant, so supply is both stable and inexpensive to transport. For these reasons, coal dominated world energy sources for a long time. However, few people today are so inflexible as to ignore established facts that stare the rest of us in the face; most people regard coal as yesterday's fuel.

Coal emissions are declining—and have been since the 1960s, when pollution issues resulted in the first effective legislation in many developed nations. Despite this decline, pollution levels remain high because of the increased use of coal by industries in previously undeveloped countries.

We have seen how coal pollutes; here, we see how it is burned in power plants to generate electricity. First, the coal is delivered to the generator site. If the coal-burning power plant is close to a mine, delivery is via conveyor belt; otherwise, it is by truck, coal train, or collier boat. Next the coal is crushed and piled in storage yards. When needed, it is pulverized, and the resulting powder is blown into a furnace. The mixture of coal dust suspended in air is highly flammable and burns at a high temperature. The furnace superheats water in a boiler—brings it up to a high temperature and pressure. The water is turned into steam, typically at a temperature of 1,000°F (540°C) and a pressure of 1,800 psi (about 120 atmospheres). The high-pressure steam is fed into turbines, causing the blades to rotate and, via the induction effect discussed in chapter 2, generate electricity. The steam that is exhausted from the turbine is then condensed via cooling towers (fig. 4.3) or by running it through pipes over which river water flows. The condensed water is then recycled back into the boiler. The efficiency of this process—the fraction of coal energy that is converted into useful electrical energy—is somewhere between 30% and 35% for conventional coal-fired power stations and will rise to about 45% for next-generation power stations that are being designed and tested today. (The main aim of these new "clean coal" plants is to reduce emissions, as we will see in the next section.)

Why bother with steam generators—why not use the water directly as input to turbines, as happens, for example, at hydroelectric sites? The answer, as so often in this field, is energy density. Steam provides energy to the turbine blades not just via pressure but also by virtue of its temperature. The

FIG. 4.3. The Ratcliffe coal-fired power plant in England. The prominent hyperboloid cooling towers, and the steam and hot air they release, dominate the site. *Thanks to Alan Zomerfeld for this image.*

steam that emerges from a steam turbine is at a lower temperature as well as a lower pressure than the steam that is input to the turbine. So steam delivers heat energy as well as mechanical energy to the turbine; the physics of steam turbines is thus described by thermodynamics, and not just mechanics (see the sidebar "Rankine Cycle").

I have already noted that burning coal in a power plant furnace generates pollutants that emerge from the smokestack—carbon dioxide, nitrogen oxides, sulfur dioxide, and water vapor, plus fly ash that contains particulate matter and toxins such as heavy metals, including mercury. You can see from my description of the operation of a coal-fired power plant that another waste product is inevitable: heat. Heat emerges from cooling towers to warm the atmosphere[21] or from cooling pipes to raise the temperature of river water. Cooling tower heat contributes to global warming by directly heating the air. The heat dumped from pipe cooling is a more immediate pollutant because it is more concentrated and can kill fish in the river. (Fish, and those who feed on them, are also harmed by mercury that accumulates within

RANKINE CYCLE

A fossil-fuel power plant is one implementation of a machine that physicists call a *heat engine*. The physics of heat engines—be they steam engines, internal combustion engines, or coal-fired power plants—are described by the Rankine cycle, named after the nineteenth-century Scottish physicist William Rankine, who analyzed the dynamics. Any engine that converts heat into work is a heat engine, and the steps in the process can be described quite generally, without reference to a particular implementation. Thus, a fluid (in our case, water) is first pressurized and then heated at constant pressure to a temperature great enough to convert the liquid into a gas (in our case, steam). The gas expands to do work (drive a turbine), causing the gas to cool and condense back into liquid. The cycle is then repeated.

The point I want to make here is that, for all heat engines, efficiency is limited. Its maximum value is $1 - T_{out} / T_{in}$, where for steam turbines T_{in} is the temperature of the steam as it enters the turbine and T_{out} is the temperature of the steam as it leaves the turbine. From this expression you can see why engineers design turbines to be driven by fluids at high input temperature and low output temperature: it maximizes efficiency. It is important to understand that in practice, other factors, such as friction, will further reduce turbine efficiency. Even if all these other factors can be made to go away (they cannot), the heat engine will still be limited to an efficiency that is below 100%, and that is determined by fluid temperatures. For practical temperatures, this maximum efficiency is about 63%.

Another result of the physics analysis is that the output power of a heat engine is proportional to the rate at which fluid passes through it. Hence (no surprise), bigger engines are more powerful. Thus, economics and thermodynamics combine to ensure that power plants that are driven by heat engines are enormous.

---

them, if the fish happen to swim in waters on which fly ash settles.)[22] The fundamental dilemma of coal—lowest energy cost and highest pollution—raises the question as to which will give ground in the future. In fact, it looks like it will be both.

A Black Future?

If we look at the figures for a typical coal-fired power plant, we can quickly determine its energy efficiency; such a calculation is provided in the appendix for the Kingston plant in Tennessee. New technologies now being developed to boost efficiency, but mostly to reduce emissions that are harmful to the environment, will be applied to coal-fired power plants within the next 15 years. The attitude of governments around the world has to be pragmatic: coal is here to stay for the next several decades. We are stuck with it, and so we must make the best of it. The improvements that the new technologies will bring to coal burning come at a price, of course. It has been estimated that the cost of coal per generated megawatt-hour of power will increase by about 40% in real terms.[23]

Steam turbines made their appearance in the 1880s. Since then, incremental increases in efficiency—such as the introduction of pulverized coal in the 1920s—have resulted in a mature technology. (There is a link between efficiency and emission reduction: recall that for every kilogram of coal that is burned, about three kilograms of CO_2 goes up the flue. Hence, an increase in coal-fired power plant efficiency results in a reduction in CO_2 emission.) The mature technology is not very efficient, quite apart from the inviolable thermodynamic reasons that we have examined. New technical advances and new incentives are needed to increase efficiency and reduce emissions. First we will look at the technological innovations, as they are the less controversial (generating less emotion and invective) and therefore easier for an engineer like me to describe.

The advances in coal-burning engineering fall into two categories—let me call them strands—with each strand consisting of several technological threads. The industry promotes these advances under the umbrella term *clean coal*.[24] The first strand is an improved coal burning method that replaces pulverized coal with *syngas*. Solid coal is converted into a combustible gas (a mixture of carbon monoxide and hydrogen) that burns much more cleanly than coal, is more efficient, and generates less greenhouse gases (read CO_2) per unit output power.[25] Other potential pollutants, such as sulfur and heavy metals, are chemically scrubbed from the syngas mixture before combustion so that they don't get spewed into the atmosphere. This new method of burning, the *integrated gasification combined cycle*, thus removes undesirable chemical elements and compounds from the coal before it is burned.

The second strand of clean coal works mostly at the other end of the combustion process, by removing fly ash and CO_2 from the emissions that result from combustion. Fly ash is removed from the combustion products by electrostatic precipitation. About half of it is recycled as road building material, or additives to cement; the remainder (which contains several toxins in trace amounts) is put into landfill.

The removal of CO_2 from emissions is referred to as CCS (*carbon capture and storage*, or *carbon capture and sequestration*).[26] Carbon dioxide that results from burning syngas is separated from the mixture of combustion products. This part of the new clean coal technology has been tested in pilot projects and works well; it removes 80%–90% of the CO_2. The CO_2 can then be disposed of underground by being injected into active or depleted oil or natural gas reserves. The geology at such sites is suitable for long-term storage of large amounts of unwanted carbon dioxide. (When CO_2 is injected into active oil reserve sites, the gas pressure can even aid in oil extraction, as we will see in the next chapter.) The process of transporting (by pipeline) millions of tons of CO_2 and burying it deep underground is energy intensive, and so CCS reduces the efficiency of coal-fired power plants and increases the cost of electricity produced by these plants. Hence, as advertised, clean coal reduces pollution but at the price of, well, higher prices.[27]

The new technologies that I have just outlined are accompanied, as part of the advocacy for clean coal, by new incentives. Both sides of the environmental debate have launched extensive and incompatible advertising campaigns over whether coal can become clean.[28] Advocates say that clean coal will be the least expensive source of electrical power in the decades to come, that the supply is secure, and therefore that coal is the most patriotic fuel, that the clean coal industry will preserve jobs and infrastructure, and that clean coal is compatible with conservation and the development of renewable sources of energy. Yet the undeniably huge efforts being made by the coal industry to clean up its image as well as its product are up against the doubters and the historical memory that the (wo)man in the street has of grimy, unhealthful coal.

Mass opinion has every bit as much inertia as a large ship: it takes a lot of effort to change course; it cannot turn around on a dime. There is a momentum to clean air legislation that is making life for clean coal advocates difficult. They are finding that clean coal is tarred with the same brush (an appropriate metaphor) as dirty old coal. Thus, for example, the World Bank is planning to restrict money to fund coal-fired power stations in the future, except for the poorest countries. At the time of writing, the U.S. Environmental Protection Agency is proposing legislation that will impose tighter regulations on emissions, ash disposal, and water usage by coal-fired power plants. These proposals are being greeted by squeals of anguish within the coal industry: they will, the industry claims, force up to 20% of power plants to close and cause a "train wreck" for business.[29]

The battle for hearts and minds, being fought out in the media as well as in boardrooms and national capitals, will no doubt continue for as long as coal is burned.[30]

The coal industry, from mine to power plant, has a bad reputation for polluting. Indeed, it does emit huge volumes of carbon dioxide, a greenhouse gas, plus particulate matter we call fly ash, plus toxins such as mercury. Elementary calculations show why so much CO_2 is produced. Coal mining has a poor safety record, particularly in emerging countries like China. Yet coal is a very popular fuel for electricity generation because it is inexpensive, has many more proven reserves than other fossil fuels, and provides a high power area density. Clean coal technology will improve coal-fired power plant efficiency somewhat and significantly reduce unwanted emissions, though the technology will take 10–20 years to become common, and then only in developed nations, because of cost.

THE SEVEN SISTERS— OLD AND NEW

In chapter 2 we saw something of the oil industry's past; here, I examine in more detail its present and future. Oil is our biggest single source of energy, as we saw in figure 2.6. Most of this energy is used for transportation, and almost all transportation is fueled by oil. Natural gas—often but not always associated with oil for geological reasons—is the world's second most significant source of fuel, and in the future it will overtake oil; between them, they account for 65% of our energy needs today. Oil and natural gas form the subject of this chapter. We will examine their extraction, refining and distribution, environmental and economic impact, and their future as primary energy sources.

Overview

We have found many applications for crude oil today beyond the obvious and most significant use as a fuel for transportation (gasoline, diesel, and jet fuel). We use oil as a lubricant; we use it as tar and asphalt. Components of crude oil make paraffin wax or can be used to manufacture different plastics.

The raw stuff that comes out of the ground is a variable mixture of different hydrocarbon compounds, as we will see when I discuss the refining process. Here, we need to know that crude oil comes out of the ground sweet or sour, light or heavy, and with a variable viscosity (resistance to flow). Sweet oil has low sulfur content. Since sulfur has to be extracted from oil before it can be used as a fuel, sweet oil is preferable to sour oil, which has a high sulfur content and is therefore more expensive to refine. Light oil has a lower density and viscosity than heavy oil. This is because it consists of a low percentage of the large, heavy hydrocarbon molecules that form resins and wax, and a high percentage of light hydrocarbon molecules, such as octane, that are used as fuel. Thus, given that most crude oil is earmarked for fuel, light crude is more desirable than heavy crude.

Oil from different parts of the world has different characteristics. West Texas Intermediate is very sweet (with a sulfur content of only 0.24%) and very light—good stuff if you are a petroleum engineer. Brent Crude (from under the North Sea off the coasts of Scotland and Norway) is sweet and light but not quite as sweet and light as WTI. Dubai Crude is light but sour (2% sulfur content). The type of crude oil that is extracted from tar sands (also know as "oil sands") is both very heavy and sour. As you might imagine, no oil company would bother with tar sands oil if there were enough sweet and light crude; 40 years ago this oil was simply too expensive to extract and refine and so was left in the ground. Today, tar sands and oil shale figure prominently in our supply and crucially in our reserves. We will see in a later section why our attitude to tar sands oil has changed.

Volatility

Here is a broad hint as to why less attractive tar sands have become important over the last 40 years: conventional oil deposits (those not from shale or tar sands) tend to be in unstable parts of the world—unstable both geologically and politically. Geologically, oil is found in young sedimentary rocks close to recently active tectonic plate boundaries—read earthquake risk. Politically, the easiest oil to extract happens to be underneath volatile parts of the human world (Iran, Nigeria, the Middle East, Russia, Venezuela).[1] The distribution of proven oil reserves is very uneven. More than half of the proven reserves are in Saudi Arabia, Kuwait, Iran, and Iraq: over 65% of crude oil comes from 1% of the giant fields.

Oil as a chemical substance is volatile. Oil geopolitics is volatile. Add to this heady mix the volatility of big business, which oil extraction and refining certainly is, and you can understand very easily why crude oil prices fluctuate wildly, a situation shown in figure 5.1, where prices are graphed for the years 1987–2011. Nor was the situation more stable in the past: figure 5.1 omits the upheavals of the 1970s (the 1973 OPEC oil embargo, the 1979 Iranian revolution). I will not provide you with a history or an analysis of the politics and economics of oil extraction. Even if I understood it (I do not),[2] it would be a lengthy distraction for a book that is about energy science, not geopolitics. We simply need to note that there is, has been since oil was first

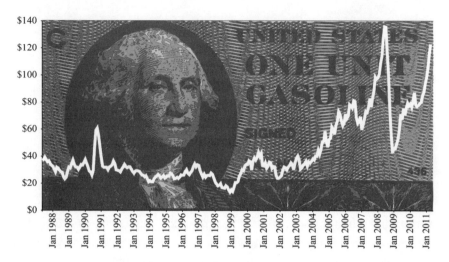

FIG. 5.1. The spot price of Brent crude oil, in April 2011 U.S. dollars, over the period 1987–2011. Over the decade from 1999 prices rose by a factor of eight, and in 2008 they fell by 60%. Long-term economic planning, both by oil-producing nations and by oil companies, is difficult in such a volatile market. *Data from the U.S. Energy Information Administration and Bureau of Labor Statistics. The background shows a 1979 gas ration coupon, never issued.*

exploited, and will be until it runs out, an inbuilt volatility to the extraction of oil, so that the oil industry is inherently cyclic. In addition, oil is an increasingly scarce resource that will probably be used up in 5–25 years.[3]

For the record, I should also note the volatility of the price of oil, which is only loosely correlated with the cost of extracting and refining. There are several reasons for these fluctuations. Shortages and excesses of oil, which drive the pump price of gasoline up and down, are manufactured for political purposes. Shutdowns caused by accidents raise prices. Most importantly, half the world's population enjoys fuel subsidies, which vary greatly from country to country. The big oil companies also enjoy subsidies and tax breaks, again widely variable among nations.[4]

The uncertainties associated with crude oil are reflected in the massive upheavals of the oil sector over the past 40 years. From World War II until the 1970s, seven large multinational companies dominated the industry. Dubbed "the Seven Sisters," these were British Petroleum (originally the Anglo-Persian Oil Company), Gulf Oil, Royal Dutch Shell, Standard Oil of California (Socal), Standard Oil of New Jersey (Esso), Standard Oil of New York (Exxon-Mobil), and Texaco (now Chevron). These Seven Sisters still exist, though much altered in some cases and with reduced influence, but they have been joined by the seven new sisters, which consist of state-owned oil and gas companies: China National Petroleum Company, Gazprom (Russia), National Iranian Oil Company, Petrobras (Brazil), PDVSA (Venezuela), Petronas (Malaysia), and Saudi Aramco. This shift in oil company power reflects a shift away from the dominance of Western private industry and toward the state industries of emerging nations.[5]

Today, 80% of proven reserves are controlled by national oil companies. The nations that run them are quite often antagonistic toward, or suspicious of, the United States and other Western countries. Yet the United States consumes more oil per capita than any other country and requires fully 25% of the world's oil production each year. These facts, plus the growing influence of emerging nations and the finite supply of crude oil, conspire to ensure that the oil industry and our dependence on it is sure to change radically over the coming decades. More on this important (indeed, crucial) subject later.

Upstream Basics

The oil industry, taken as a whole, is the world's largest. That is, it spends and reaps more money than any other. It is so big because it is so important (for example, in the United States alone it employs 1.8 million people). Being big, it is also powerful and attracts heavy subsidies and tax breaks in the United States and other oil-producing and -consuming nations. Each year, humanity sucks out of the earth, refines, and burns over a cubic mile of crude oil (30 billion barrels). Such a large industry has become very specialized, as you might imagine. The oil industry divides the different aspects of oil work into two broad categories: upstream work, which is everything to do with the exploration for and production of oil, and downstream work, which encompasses all the rest. (The whole cycle, from exploration to abandonment of an exhausted oil field, is known within the industry as the *value chain*.) In this section we look at the upstream oil industry.

EXPLORATION

Prospecting for oil is a risky venture, like prospecting for gold. Also like gold prospecting (and other mining ventures), the search for oil uses techniques that have evolved from something akin to black magic into a highly specialized scientific discipline. Geologists, geophysicists, engineers, and armies of technicians are involved. They conduct geological surveys to generate as much information as they can about the formations that lie below ground, and then conduct remote sensing tests and analyses to provide more detail. Indeed, today's petroleum geologists are able to create three-dimensional images of the strata beneath the surface. It used to be the case that oil seeped to the surface, giving itself away. All such accommodating oil reservoirs have long since been detected and used up; today it is much harder to find oil.

Gravity surveys provide information about the different densities of rock formations below ground, and magnetic surveys tell petroleum engineers about mineral content. Petroleum geologists know from long experience that some formations and certain minerals are associated with oil reservoirs. Thus, for example, oil develops in rocks that once formed the beds of ancient seas; salt domes (underground strata of salt) also result from such seas, and so the presence of a salt dome might indicate an underlying oil reservoir.

Core samples provide physical evidence of what lies underground. Electronic sniffers provide further clues. More sophisticated remote-sensing techniques include seismic reflection surveys. Such surveys are undertaken by geophysicists, who set off explosions underground or in seas, sending acoustic waves through the earth's core and mantle; these waves reflect off formation surfaces (in particular, they reflect off boundaries between strata of different density), and the detected reflections are processed by computers to yield 3-D images. Some seismic information can be obtained by *thumper trucks*—mobile sound sources that send sound vibrations into the ground.[6] Forming images of the earth's structure below the surface by sending acoustic waves underground is closely analogous to medical ultrasound imaging, whereby acoustic transmitters placed on the skin send sound waves into tissue. Reflections are then detected and processed to provide a picture of your unborn child or a kidney stone. It was just such new technology that led to ExxonMobil's discovery of very large oil reservoirs under the Gulf of Mexico in 2011.[7]

Via this panoply of sophisticated measuring techniques, plus old-fashioned geological fieldwork and a mass of analysis, oil reservoirs are located. From the reservoir depth and surrounding geology, petroleum engineers can form a pretty good idea of the size of the field and the type of crude it contains. If the oil is considered to be economically recoverable (accessible, of good enough quality, in large enough amounts), then drilling begins.

DRILLING

As you may imagine, constructing a well bore several miles in length is far from trivial. The techniques and technology have been greatly refined and improved during the last century. The historical method involves drilling a vertical bore and inserting a cylindrical steel casing in sections to prevent infill. The diameter of the casing varies from well to well but can be as much as 20 inches (half a meter). The casing is cemented in place. A *drill string* (a long rotating tube with a complicated and very hard drill bit at the end) is then passed down the pipe. Drilling fluid (of which there are several types, all generally referred to as *mud*) is pumped through the drill bit and brought back to the surface via the casing, along with debris from the drilling process (broken rock, etc.).

Modern drilling permits a direction change at a desired depth. That is, the borehole piercing the ground does not have to follow a straight line. This is a neat trick—technically difficult to achieve and consequently one that has developed only over recent decades. The easiest way to explain how direction change is achieved is by pointing you in the direction of a YouTube video (a picture, especially if it is moving, is certainly worth a thousand words in this case).[8] The key component is a *whipstock* casing, which deflects the drill bit to a controllable degree. *Directional drilling* has several advantages that justify its greater expense (by a factor of two or three) compared with drilling in a strictly vertical direction:

- More well heads are able to access a particular reservoir (imagine several straws dipped into the same milkshake).
- It becomes possible to drill around, rather that through, a difficult formation (a bendy straw goes around a scoop of ice cream instead of through it) or to drill under a lake from a nearby onshore rig.
- The reservoir area accessible from a single wellhead is increased by a factor of about 20.

This last advantage—increasing the area of reservoir that can be reached—applies for the following reasons. Oil reservoirs are not usually bodies of liquid, like underground lakes. Often they consist of porous rock saturated with oil droplets. To release these droplets the rock must first be broken up, or at least be close to the bore hole, so the oil can ooze out. Oil-bearing rocks are often in horizontal strata; thus, a directional drill, which can switch from vertical to horizontal at the depth of the oil, is very productive because it can drill *along* the stratum and not bore right through it. (a horizontal bore can tap much more oil than a vertical one if the stratum is thin.)

Needless to say, all this exploration and drilling is extremely expensive, particularly if the oil is offshore. For an offshore rig, 65% of its total costs are up front, while the running costs can go up to $464,000 per day for a deepwater rig (much less for an onshore rig).[9] These expenses exclude the cost of cleanup for any spills that occur (more precisely, perhaps, spills that are seen to occur). Clearly, only very large companies with very deep pockets can play this game.

EXTRACTION

Oil from a new well really can gush, just like in the movies. A conventional oil reservoir (one tapping into liquid oil) may be under natural pressure which, when the reservoir is penetrated by a drill, forces oil up to the surface. On average, something like 10% of the oil that is in a reservoir is pushed out by natural pressure. This stage is called *primary recovery*. Before long, the pressure drops, and more active methods are needed to extract oil from the well. Pump jacks (the familiar "nodding donkeys" seen across many U.S. oil states) may be employed to bring oil to the surface. During the *secondary recovery* stage, water is injected into the well under high pressure, or natural gas that emerges is collected and reinjected under pressure, thus forcing oil out of the reservoir. This stage recovers perhaps 30% of the oil. The final stage is *tertiary recovery*, which consists of injecting steam or hot CO_2 at high pressure into the well. This action heats the oil, reducing its viscosity and so helping it to flow. Perhaps another 10% of the reservoir is brought to the surface during the tertiary stage of extraction.

Oil that is bound up in porous rock requires further encouragement before it can be extracted. In this case the high-pressure water (usually brine) that is injected serves to fracture the porous rock and release the oil. This hydraulic fracturing (*fracking*, in the trade vernacular) of the rock is achieved via *injection wells*. Consider the common case in which many oil wells tap into a single large reservoir. A carefully chosen pattern of injection wells—they are often old production wells—will frack the oil-bearing rocks and push the released oil toward a production well, which brings it to the surface. (The water used for fracking contains other chemicals, such as hydrochloric acid, which environmentalists fear can leach into groundwater.)[10]

Downstream Work

The downstream sectors of the oil and natural gas industries include refining and distribution. The main components of interest to us deal with the transporting and refining of crude oil.

TRANSPORTING

Within national boundaries oil is generally moved around by pipeline. The crude emerges from a production well and is piped to a port or refinery. For longer journeys oil tankers provide the means of transportation. Both these carriers are efficient (costing a cent or two per gallon to move from A to B, wherever A and B might be), both date back to the second half of the nineteenth century, and both are susceptible to accidental or intentional disruption—of which, more anon. Other carriers (trucks, trains) are more expensive and therefore constitute a much smaller slice of the transportation pie. In this section I concentrate upon pipelines and supertankers.

For some sources of supply, pipelines and oil tankers both play a role. For example, the Trans-Alaska Pipeline System (TAPS—see fig. 5.2) carries oil from production wells around Prudhoe Bay on the North Slope of Alaska across the state to Valdez, an ice-free port on the southern coast, where the oil is poured into tankers. These tankers then convey the oil to the rest of the world, barring accidents. Another instance using dual forms of transportation is the SUMED (Suez-Mediterranean) pipeline, which runs from the Mediterranean coast of Egypt to the Red Sea, bypassing the Suez Canal (which is too narrow for the very large modern supertankers, as we will see).

In the United States some 68% of domestic oil is moved around the country by pipeline, compared with 27% by tanker. Clearly, there are economic and geographical factors that influence these numbers, and so they vary from country to country, but they serve to show the relative importance of pipelines.[11] The same geography, plus politics, explains why America has 10 times more oil pipeline (165,000 miles) than does Europe. A little less obviously, the volume of oil transported by pipeline is twice the annual consumption. (A given gallon of oil is typically moved twice—once to a refinery and a second time from refinery to distributor.)

Both pipeline engineering and pipeline business infrastructure are complex and sophisticated. Some of the engineering considerations are discussed, with TAPS as the example, in the appendix. For instance, the TAPS pipeline is placed on sliders (see fig. 5.2), to allow for thermal expansion and contraction and

FIG. 5.2. A section of the Trans-Alaska Pipeline System at the point where it passes over the Denali Fault. To reduce the likelihood of the pipe breaking during earthquake movement (or, indeed, as a result of thermal expansion) the pipeline is placed on sliders. *U.S. Geological Survey image.*

accommodate earth tremors. The aboveground sections of TAPS (much is underground) must cross rivers and be crossed by herds of migrating caribou. All sections must pump warm oil hundreds of miles, cross mountain ranges, and detect and notify pipeline operators about leaks.[12] Pipeline companies are separate entities from oil producers; they do not own what they ship. Often, different products move down a pipeline at the same time, like a train with railroad cars carrying different loads. The conveying of these different cargoes around a large network of pipelines so that a customer can move his product from *A* to *B* in an agreed time and at an agreed price is a complicated task, as you may imagine.

The volume of crude oil transported each year by sea in tankers increased from about 1.1 billion tons in 1970 to 1.9 billion tons by 2005, and is currently over 2 billion tons per annum. In 2005 a quarter of all the world's tankers carried crude oil, and 62% of the world's oil was conveyed by tanker. The size of the oil tankers has increased tremendously since 1970; today, a single large supertanker can carry over half a million tons of crude.[13] The cost, at two or three cents per gallon, is greater than the cost of transporting by pipeline but still is very low.

Changes have also been made to the regulation of tankers since the 1970s. It used to be the case that a tanker, which under international law must be registered with a particular nation (the so-called "flag state"), could fly a flag of convenience. That is, it could fly a Panamanian or Liberian flag or the flag of some other country with lax safety or pollution regulations. Lack of regulation means (in oil transport as well as banking) lower operating costs and greater profits but also increased risk. Crew injuries and environmental damage are both more likely if conditions of employment or working conditions are slack. Several widely publicized accidents at sea—widely publicized because of the large amounts of oil spilled; large amounts of oil spilled because of large tanker sizes—have led to more stringent regulations that reduce the likelihood of serious spillage.

Reduce but not eliminate. Oil spillage is a major factor in the minds of many people and raises the question of the vulnerability of oil transportation. I will discuss accidental spillages, the environmental damage they cause, and the cost of cleanup later in this chapter. Here, let us consider the vulnerability of oil transport—by tanker or pipeline—to intentional disruption.

VULNERABILITY

In 2001 a drunkard with a hunting rifle punctured the Trans-Alaska Pipeline with a single bullet. The bullet penetrated the half-inch steel exterior of the pipe and a 4-inch layer of thermal insulation inside it (recall that transported crude is kept warm to reduce friction). This single incident resulted in 150,000 gallons of crude being sprayed onto the Alaskan tundra. Pipeline workers could not move in to fix the leak until the area was declared safe, so the pipeline was shut down for several hours. Production of oil on the North Slope had to be reduced by 5% during the shutdown.[14]

One bullet, one stupid act that took a second to execute, and a pipeline that carries one-sixth of the U.S. domestic oil supply was shut down. The cost and effort required to fix the damage was out of all proportion to the cost and effort of causing it. It is impossible to protect an 800-mile pipeline —an obvious fact that was widely reported at the time. This bullet was not fired by a terrorist, but the incident shows that willpower and motivation are all that are needed (apart from a few well-placed bullets) to cause huge disruption and damage to a nation's pipelines.[15]

In less stable parts of the world, oil pipelines simply shut down, and for extended periods, during conflicts. Thus, the Iraq–Saudi Arabia oil pipeline was closed during the first Gulf War. A Libyan pipeline carrying sweet, light crude was shut down in 2011 at the beginning of the recent revolution there. Pipelines cannot be relocated to safe positions during wartime, they cannot be protected along their entire length, and one breach at any point serves to sever the supply. Faced with these hard facts, the only way to prevent an enemy from shutting down your pipeline is to shut it down yourself.[16]

Risky though pipeline supplies are, they are considered more reliable, and perhaps no more vulnerable, than tanker supplies. Tankers may originate in or pass through politically unstable or hostile regions, and they are individually very much at risk of attack. Thus, the U.S. Government Accountability Office—and no doubt many other departments—maps maritime chokepoints whose safe passage is critical to the supply of crude oil.[17] These show that the densest oil tanker traffic is through the Strait of Hormuz at the mouth of the Persian Gulf. The busiest sea lane is from there to the Strait of Malacca, which runs between Malaysia and Sumatra, and through which the oil-thirsty Far East is fueled. Over 300 million tons of oil are transported along this sea lane each year. Other hot spots are the Strait of Gibraltar, the Cape of Good Hope, and the Straits of Florida, through each of which over 200 million tons of oil pass annually. Clearly, any pirates, terrorists, or regional governments that felt the need to pick off supertankers could very quickly cause chaos and shortages around the world. Such action would be expensive to counter and would ramp up political tensions.[18]

Viewed on a larger scale—independent of the means of transportation— our supply of oil could be greatly reduced, with consequential price hikes, by foreign governments hostile to our interest or policies. A kerfuffle in Kuwait or dubious diplomacy in Dubai may put up the price of petroleum in Pennsylvania. Half of U.S. oil is imported, and this dependence on foreign oil gives oil-producing countries leverage over the United States. As a consequence, American foreign policy must take heed of oil supply considerations —a fact brought home 40 years ago as a result of the OPEC oil embargo of 1973, but one that has influenced foreign policy for much longer. I do not want to get into the merits and demerits of Western governments' policies and actions over oil, whether political, diplomatic, or military, but will merely state the bald fact that our supply of oil is partly out of our hands. Our industry currently depends sensitively upon oil; it can be severely curtailed or entirely shut down for days or weeks by a decision made on the other side of the world.

It *may* be possible to overstate the importance of oil to our national well-being—certainly, many people have tried—but on the other hand, it may not. Western nations are adopting a number of policies to reduce the consequences of a disruption to oil supply. For short-term problems (that dubious diplomacy in Dubai), stockpiling may smooth out a temporary oil drought. The most important and most obvious is to reduce our dependence upon oil. The geopolitics of oil and the dwindling world supply work together in this instance to motivate our development of alternative energy sources. Another long-term policy that is already being acted upon is to do what we can to secure supplies from friendly or reliable producers. This policy alone may account for the huge investment made over the past three decades in the development of Canadian tar sands: the United States now imports twice as much oil from Canada (a quarter of all imported crude oil) as from Saudi Arabia.[19]

REFINING

What is this messy black liquid that comes out of the ground? Crude oil consists mostly of hydrocarbons—molecules made up of carbon and hydrogen atoms. There are hundreds of hydrocarbons, and the mixture that constitutes crude oil varies from region to region.

The geometrical arrangement of carbon and hydrogen atoms determines the physical and chemical properties of the molecule. The three hydrocarbons illustrated in figure 5.3a, each with a backbone of six carbon atoms, illustrate the kind of variations that exist. (The name *hexane* reflects the number of carbon atoms.) Hydrocarbon chain molecules (*alkanes*) with 1 to 4 carbon atoms yield gases such as methane (CH_4), ethane (C_2H_6), propane (C_3H_8), and butane (C_4H_{10}).[20] Alkanes with 5 to 8 carbon atoms form constituents of liquid gasoline. One of these is octane, with 8 carbon atoms, which is of course a key ingredient in gasoline. Heavier alkanes with 9 to 16 carbon atoms are used for diesel and jet fuel and kerosene. Longer chains correspond to solids: the alkane with 25 carbon atoms is paraffin wax; those with more than 35 carbon atoms make up asphalt.

As we have seen, other atoms (such as sulfur) are present in crude oil, but these are impurities that must be removed by refining the crude. Refining also separates the crude oil into its constituent hydrocarbons, such as butane

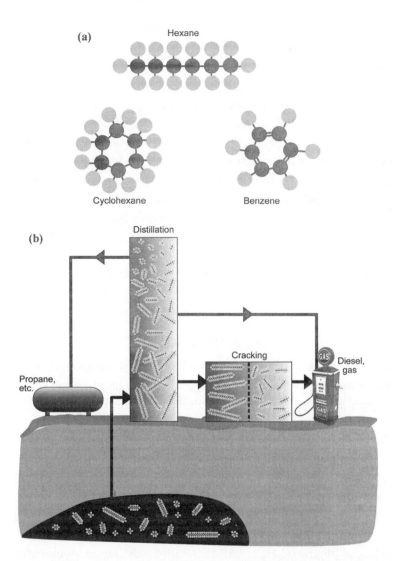

FIG. 5.3. Hydrocarbons and hydrocarbon refining. (a) Many types of molecules can be constructed from only carbon and hydrogen atoms. Shown are representatives of three families containing six carbon atoms. The chain molecules (alkanes) are of most interest to us, as these molecules provide us with our liquid and gaseous fossil fuels. The short lines indicate chemical bonds; note that hydrogen atoms each have one bond and carbon atoms have four bonds. (b) An oil refinery schematic. The mixture of alkanes that come out of the ground are separated by distillation. Some of the heavier molecules are cracked (the carbon chains are broken) to yield smaller alkanes that are more useful.

and octane, and splits up (*cracks*) the heavier long-chain hydrocarbons into lighter, more useful ones. These refined petroleum products are then transported to distribution centers and sold to you and me, to power our industries, cars, barbeques, or cigarette lighters. A schematic illustration of refinery processing is shown in figure 5.3b.

Not even its most passionate advocate could describe an oil refinery (fig. 5.4) as pretty. It does a messy job with messy products and byproducts and cannot be dressed up to blend in picturesquely with its environment (unlike, say, a nuclear power station). In this respect oil is similar to coal; it is unavoidably ugly.

One barrel of crude, 42 U.S. gallons, is refined to produce about 44 U.S. gallons of petroleum products: the refining process generally reduces the density of the molecular constituents of crude oil, thus increasing their volume. Roughly half of each barrel of crude ends up as gasoline, a quarter as diesel, and 10% as jet fuel. Gasoline consumption in the United States has

FIG. 5.4. An oil refinery near Anacortes in Washington State. Perhaps the breathtaking scenery only adds to the ugliness. This refinery produces 145,000 barrels per day of refined petroleum products—gasoline, diesel oil, jet fuel, propane, and coke. A blaze in April 2010 killed four people and critically injured three others; production capacity was cut to 70% of normal during cleanup. *Photo by Walter Siegmund.*

increased by 60% since 1970; Americans now guzzle 43% of the world's annual production.

Peak Oil

Fossil fuels are finite resources. There will come a year for each one—coal, oil, natural gas—when the amount produced reaches a peak. We have seen that the year for peak coal will arrive sometime in the next three decades. What about peak oil? Oil production increased from zero in the nineteenth century; it will cease entirely sometime in the twenty-first when the supply runs out; ergo there must be a year of maximum production. Clearly the date is important; because of the predominance of oil as a fuel source, it indicates roughly how much time we have to produce an alternative source of energy. The trouble is that experts differ considerably in their predictions for peak oil; some say that it has already passed.[21]

The emerging markets, particularly the skyrocketing growth of China and India, coupled with the still-expanding use of gasoline by the world's main consumer nation ensure that, without an alternative energy source, we will experience a hard landing. U.S. oil production peaked in the 1960s. Here is just a faint hint of desperation: while we still have 350,000 oil wells, more than one-third of these produce less than *one barrel* of crude per day. Indeed, 79% of them produce fewer than 10 barrels a day.[22]

Given that oil is our primary source of fuel, if behooves us to ask when the tap will run dry. The CEO of Shell, Jeroen van der Veer, said in 2008: "Shell estimates that after 2015 supplies of easy-to-access oil and gas will no longer keep up with demand." On the other hand, Guy Caruso of the U.S. Energy Information Agency said in 2005: "A peak in world oil production is decades away . . . not years away."[23] Here we have the production prediction problem in an oil barrel: peak oil is years away, it is soon, it has already happened. There are many such predictions, many sound bites and column inches of press reporting, and the inevitable conspiracy theories telling us that the government is doctoring the figures.

I asked David Greene of Oak Ridge National Laboratory and the University of Tennessee about the accuracy of peak oil predictions. Greene has 30 years' experience in the field, and he says that the predictions are getting pretty sophisticated and reliable; they have come a long way since the first simplistic statistical analysis of the 1950s (the *Hubbert curve*). Predictions can now be made for individual oil sources, not just a broad-brush estimate for the planet as a whole. He provides an example: the International Energy Agency prediction of 2010 for non-OPEC peak oil seems to be borne out by recent data.[24]

Despite the improvements in prediction accuracy, we have seen that the predictions for peak oil vary widely from one organization or government to the next. Partly this is due to the adoption of different measures, and no doubt it is also partly due to bias and dissimulation. For what it is worth, neglecting outliers, the figures widely quoted and easily found by Googling "peak oil" predict that oil production will peak sometime between now and 50 years hence. Predictions by U.S. government organizations tend to be the most optimistic (placing peak oil furthest in the future), private oil company predictions less optimistic, and academic institution predictions more pessimistic.

Maybe the different predictions arise because they are based on data from different parts of the world. Maybe they arise from factors other than hard science: different groups with different interests are drawn towards an earlier or a later date for peak oil. Another source of confusion is the fact that peak oil predictions usually are restricted to conventional oil, explicitly excluding oil from shale and tar sands (which I will discuss soon enough). There are signs within the industry that suggest to me the imminence of peak production, at least for conventional oil.[25] Here is a suggestive though perhaps too emotive analogy. Consider the large oil companies to be a pack of hungry wolves and the oil reserves of Mother Earth to be a caribou that they are tearing apart and eating. Every last piece is devoured hungrily—except for a few scraps that are snapped up by smaller predators such as coyotes. The coyotes also pick over the bones that remain after the wolf pack has moved on to another kill.

These coyotes correspond to the many small oil engineering and production companies that depend upon the big companies for their living—providing infrastructure, maintenance, and manpower as well as specialist expertise. Alan MacFadzean worked for many years for such a coyote company in the North Sea sector. His view is that these companies are already suffering from the consequences of reduced oil production and slowed discovery of new reserves.[26] Inevitably, the coyote companies will feel the pinch before the wolf companies because the wolves are big enough to survive periods of starvation. Also, as apex predators they call the shots.

What do the big oil companies themselves think about peak oil and the subsequent crunch? I turned to Peter Ward, a physicist with 30 years' experience in the field who works for one of the original Seven Sisters—a wolf company (though nowadays his company's reserves are less than those of Petronas, the Malaysian state oil giant). He points out that there are several definitions of peak oil and that they are not the most important measure of our finite oil resource. A better measure is the year in which demand exceeds maximum possible supply (the supply available when all producers decide they want to extract as much oil as they possibly can, with the supply not constrained by political or price-fixing considerations). In Ward's view, this will happen within the next 10 years. He cites the largest conventional oil field ever discovered—Ghawar in Saudi Arabia, which generates one-seventh of the world's oil: it now produces 30% water. This fraction increases from zero as a field is drained (injected water is part of the secondary phase of oil recovery, you may recall). Ghawar will continue to produce for many years to come—only when the water content exceeds 80% will it be uneconomic—but the writing is on the wall.[27]

Of course, oil companies continue to look for new reserves. It is their single biggest concern, and yet despite the new technology and their best efforts, the reserves of all oil companies are shrinking. (Today the hot spots for conventional oil exploration are offshore Brazil and offshore Angola.) On the supply side, the peak year for oil reserve discoveries was sometime in the 1950s; on the demand side the peak year is yet to come—the massive hike in demand from China and India was not anticipated 30 years ago. The combination of these factors, plus the undoubted depletion of conventional reserves, means that the crunch is coming soon.

The United States can, in the view of my experts, become self-sufficient in energy (and thus independent of the whims of foreign suppliers who do not necessarily have the best interests of the United States foremost in their thoughts). Tar sands and shale may not be the answer—a more efficient, simpler, and more environmentally friendly way would be to reduce U.S. per capita demand to that of Western Europeans.

Big oil companies do not care about the environment or about what the public thinks of them (except as the environment and public opinion influence government policy). Here is an illustrative demonstration. Contrast the reaction of the media, and of governments (and hence the oil companies), to two recent spills. A huge spill at the BP-Macondo prospect in the Gulf of Mexico (the infamous Deepwater Horizon disaster, which I describe in more detail later) generated a massive cleanup operation, thousands of lawsuits, and billions of dollars of costs and fines. The recent Chevron Endeavor blow-out is much less reported. An explosion in January 2012 at a Chevron natural gas rig 10 km off the Nigerian coast killed two workers and caused fires that burned for weeks, eventually causing the rig to collapse into the Atlantic Ocean. Local communities suffered from pollution and disruption to fishing. The scale of the Chevron disaster was smaller than the BP disaster, but the media response (and that of the Nigerian government and Chevron) was disproportionately smaller. Had the Chevron Endeavor explosion occurred in the North Sea or the Gulf of Mexico, we would have heard much more about it.[28]

Big oil companies care about reserves and about oil prices. High prices make them optimistic, and so they invest (hugely) in exploration and the search for more reserves. Low prices make them pessimistic, and they withhold investment and cut costs. Given the long interval between investment in exploration on the one hand and production on the other, oil companies are very wary of volatile fluctuations in the price of oil, such as we have seen over the past forty years.

Scraping the Barrel

To continue an earlier analogy: the wolves have eaten all the tender flesh and are now obliged to feed upon less-appetizing gristle and bone. ExxonMobil spokesman William J. Cummings explains the situation more plainly, in a widely quoted prediction: "All the easy oil and gas in the world has pretty much been found. Now comes the harder work in finding and producing oil from more challenging environments and work areas."[29] Gone are the days when oil gushes out of the ground in our own backyard. We next found it in somebody else's back yard, and then we found it offshore. Now we find it in shale and tar sands. These are increasingly difficult places from which to extract oil: crude from gushers in California, Pennsylvania, or Texas costs less per barrel to extract than crude from Kuwait or Saudi Arabia or Venezuela. Oil beneath foreign soil costs less to extract than oil from beneath the North Sea or the Gulf of Mexico (because of the technological challenges of drilling through the ocean bed from a rig on the ocean surface). Offshore oil costs less to extract than oil from shale or tar sands, for reasons that will soon be made clear. The increased costs become economical only in an environment of high oil prices; the imminence of peak oil and the subsequent crash in oil supply is driving prices upward—like food prices in a besieged and starving city. Hence the current and future interest in oil shale and tar sands.

SHALE

Oil shale is a porous rock containing *kerogen*, which is a mix of hydrocarbons formed from partially decomposed algal material. Kerogen is bound to the shale and is a crude oil precursor. That is, the complex series of geological processes that lead from organic matter to oil also lead to kerogen; one final stage (baking at high pressure and temperature) would be enough to turn the kerogen into crude oil. Kerogen can be separated from oil shale, and tarlike *bitumen* extracted. Bitumen is basically asphalt, and from it we can derive octane, butane, and the other oils which fuel our lives.

From this brief description, you can see that getting oil from shale is more complex than extracting it from crude, because nature has bound the kerogen to shale rock—it can be released by heating it to 400°C–500°C—and has stopped one stage short of turning it into crude oil. Thus, the cost of extracting shale oil (known as *synthetic crude* because of all the processing that it requires) is much higher than that of extracting natural crude oil—not least because it consumes 70 units of energy (usually from associated natural gas) to create 100 units of petroleum energy. Indeed, extracting shale oil cannot yet be done on a large scale in a manner that is economically worthwhile. Also, the conversion process generates a great deal of carbon dioxide, and so shale oil is doubly problematic for the environment: CO_2 is generated when we make it and again when we burn it.

Note the difference between *oil shale* and *shale oil*. Oil shale is the rock, containing kerogen. It is mined in open cast or underground mines, in a manner similar to coal mining; and indeed, oil shale can be burned directly like coal. Shale oil is the synthetic crude derived from kerogen. Sometimes conventional crude oil is found in oil shale formations: this is known as *tight oil*. Natural gas is also found in oil shale; extraction of this *shale gas* is easier than extraction of shale oil because the natural gas is not bound to the rock.

It is not difficult to understand why oil shale has been left in the ground, given that conventional crude oil is less expensive to extract and less environmentally damaging. By the same token, it is not difficult to see why oil shale will be an increasingly significant part of our energy future, as conventional oil peaks and then declines. Since the 1970s the exploitation of oil shale has increased—intermittently, depending upon crude prices. Here is the one great advantage of oil shale: there are huge volumes of the stuff not very deep underground, and a lot of it is close to home. In fact, the enormous Green River formation in Colorado, Utah, and Wyoming, is mostly on federal land.

In 2011 the U.S. Geological Survey estimated that this formation contained 1.44 trillion barrels of oil, though this estimate made no attempt to assess how much of the oil was recoverable because currently there is no economically feasible extraction method. Later estimates put the volume of recoverable shale oil at 800 billion barrels, conservatively. This number is three times the volume of Saudi Arabian conventional crude reserves. There are other large deposits under the North Slope in Alaska, and in Canada. Clearly, if an inexpensive means can be found to generate synthetic crude from kerogen, then peak oil can be pushed backed many decades.

Perhaps it is needless to say that the exploitation of oil shale is politically divisive. One side says it is an energy necessity without which the United States and other developed countries will become impoverished or worse (*we're doomed if we don't!*). The other side says that we would be destroying our own environment if we do exploit shale, and so it is best left in the ground (*we're doomed if we do!*).[30]

TAR SANDS

Like shale, tar sands contain unconventional oil that is expensive and messy to exploit. Also like shale, tar sands hold massive amounts of oil and are close to home. Unlike shale, tar sands yield their deposits today: oil extraction is currently feasible. Given the scarcity of new conventional oil sources, tar sands are currently being exploited on a large scale.

One of the biggest oil deposits in the world lies under northern Alberta, Canada. The Athabasca tar sands contain 174 billion barrels of proven oil reserves in the form of bitumen, the asphalt-like tar that is too viscous to flow like oil. It is processed to yield synthetic crude, as we saw with oil shale. The energy density of tar sands (the amount of oil per cubic meter) is only 10% or 20% that of oil shale, and so the Athabasca deposit, though it is smaller than the Green River formation, requires a larger area to yield each ton of oil. Because some 20% of the tar sands are within 200 feet of the surface, the sands can be strip-mined. The sands are scooped out of the ground, hot water is added, and the bitumen is separated from the sand and processed. The remaining sand and water slurry is deposited in large tailings ponds, where the sand settles out as sediment and the water (95% of it, according to the Canadian Association of Petroleum Producers) is recycled.

The deeper tar sands—more than 200 feet below ground—are extracted in a very different manner. Two pipes are placed within the underground sand deposit, one above the other. The upper pipe feeds steam into the sands, which heats the bitumen and causes it to become more fluid. The bitumen flows down to the lower pipe, from which it is brought up to the surface, along with the condensed steam. It is then processed to form synthetic crude. This underground extraction process is called SAGD (steam-assisted gravity drainage). It has a much smaller surface footprint than the strip mining method of bitumen extraction but is, of course, more expensive.

Bitumen extraction in Canada's tar sands is controversial because of the environmental and health issues that arise, particularly from the strip-mining method. Each cubic meter of oil that is produced from the Athabasca deposit requires 2–4 m³ of water, which is taken from the Athabasca River. The allowed amount is low (only 1% of the flow), but environmentalists worry about the consequences for river life and for people downstream of the tar sands, particularly in winter, when the water flow is lower. Although recycling the water used in strip mining saves water, it also tends to concentrate naturally occurring toxins such as mercury, arsenic, and lead in the tailings ponds. There are questions about fish deformities in Lake Athabasca and about elevated cancer risks for people who live in Fort Chipewyan, downstream of the tar sand strip mining. The mining industries claim that they are heavily regulated and that their environmental and health impact is small. Both they and their opponents have become expert at reaching out to the public with research papers, newspaper reports, and videos. The issues are not likely to go away soon.

Bitumen mining in the Athabasca tar sands generates large numbers of jobs—perhaps 100,000 in Canada alone. Some 80,000 American jobs also depend upon these tar sands, and it has been claimed by proponents that another 80,000 will be created if the Keystone pipeline is extended. This pipeline transports the synthetic crude to various locations in the United States, but in November 2011, President Obama delayed its second phase, the Keystone XL extension, until 2013, to the delight of environmentalists and the chagrin of the petroleum industry. There are wider geopolitical issues involved in the transportation of Athabascan synthetic crude. The Keystone pipeline divides both Americans and Canadians along political as well as environmental lines. If the Keystone extension is rejected, the alternative is to build a pipeline (the Enbridge Northern Gateway Pipelines Project) to the Canadian Pacific coast for exporting the crude to Asia. This alternative has also generated opposition from environmentalists.[31]

It's a Gas

Across the globe, we currently consume about 31 billion barrels of oil each year. Given the estimated reserves of unconventional oil, the Athabasca tar sands can supply the world for about five years and the Green River shale formation for maybe 25 years—assuming no future increase in consumption. It is very likely that consumption will increase, and so these supply numbers are optimistic. Shale and tar sands can delay peak oil but only for a matter of decades. What about that other significant fossil fuel, natural gas?

Natural gas is a sibling to oil, and therefore much of what I have said about oil applies to natural gas. Consequently, in this section I will emphasize the differences. We have seen that natural gas can be associated with oil—found in the same well—or it can be isolated ("non-associated natural gas"). Natural gas can be compressed or liquefied, and it is transported by pipeline or by tanker, as is oil. I will say something about gas pipelines, because they are not quite the same as oil pipelines. The main differences between these siblings—differences that are important for our future—are that gas is now cheaper than oil (per unit energy extracted), is more plentiful (in terms of remaining reserves), and burns more cleanly. Another difference, of geopolitical importance, is that the distribution of natural gas reserves across the nations of the world is not quite the same as the distribution of oil reserves; the largest gas field is off the coast of Qatar (the largest oil field is in Saudi Arabia), while the largest gas reserve is in Russia (the largest oil reserves are in the Middle East).

If no new conventional oil is found and if the oil consumption rate stays the same as it is today, we will run out of oil in 42 years. The equivalent figure for gas is 60 years. In energy terms the proven oil and gas reserves are about the same. Geologists estimate that the amount of recoverable oil yet to be discovered in the world equals about half the current level of oil reserves, whereas about 70% of natural gas reserves are yet to be found. I do not know how much faith to put in these figures, but the suggestion is that gas is more plentiful.

Natural gas is about 80% methane and 20% ethane, with trace amounts of propane, butane, pentane, and other hydrocarbons, plus contaminants such as sulfur. The raw product is processed to remove contaminants and condensed liquids (the gas expands and cools as it is brought to the surface, and so tends to condense), and then the different hydrocarbon constituents are separated. The natural gas we burn is pure methane, which burns cleanly. It also generates less CO_2, per unit of energy released when burned, than do oil or coal. But "less" is relative here: natural gas is a fossil fuel and generates CO_2 in large amounts. Moreover, it is itself a greenhouse gas, and so leaks of natural gas are environmentally damaging.

Natural gas is a relatively reliable source of fuel because it is piped directly to its domestic users, whereas oil and coal are brought in trucks, weather permitting. Gas pipelines are divided into three functions: gathering, transporting, and distributing. Gathering pipes tend to be small-diameter and low-pressure; they take gas from individual wellheads to a processing plant. Any sour gas (gas with a high sulfur content) that is produced must be transported in separate pipes because it is corrosive. Distribution pipes are also small in diameter (typically a half inch for domestic distribution).

The pipes used for long-distance transportation of natural gas are large (16–48 inches in diameter) and move gas at higher pressures. The pressure is maintained along the length of the pipe by compressor stations every 40 to 100 miles. At these stations, any water or hydrocarbon condensate found in the pipe is removed. Valves permit the flow to be directed, and facilitate repairs and replacement of pipes. The pipes themselves are made from steel and are joined together in such as way as to follow the land contours. The pipeline, welded airtight, is sealed with a coating that protects against corrosion. Most long-distance pipes are underground. Control stations monitor the flow rate, temperature and pressure, permitting real-time adjustments when necessary. As with oil pipelines, the interior of natural gas pipelines is

---

HOW FAR ON A TANK OF GAS?

A quick calculation will provide us with an estimate. In the appendix I use simple physics to show that, for a small car of specified aerodynamic drag and fuel capacity, a tank of gasoline will move the car a distance of 680 km. The same car powered by natural gas, with the same fuel capacity (assuming that the natural gas is compressed, at 160–200 atmospheres), will travel 210 km on a tank of fuel. The difference is due to the lower energy density of compressed natural gas (CNG) compared with gasoline. The high pressure also means that a CNG tank must be much more robust than a normal gasoline tank. Both these reasons argue for a *large* CNG tank—and this is why, at the time of writing, CNG-powered vehicles tend to be big (trucks and buses).

---

inspected and cleaned by pipeline inspection gauges. There are over 300,000 miles of natural gas pipeline in the United States.

A recent and detailed study by the Massachusetts Institute of Technology considered the future of natural gas as an energy source for the United States. The authors concluded that natural gas is abundant, clean, and flexible and sets the benchmark for fossil-fuel CO_2 emissions. It is "finding its place at the heart of the energy discussion." Natural gas will be increasingly used in the transport sector as gasoline becomes scarcer and more expensive.[32] Gas already provides 23% of American electrical energy, and its use is increasing; production spiked 7.4% in 2011.

Cleaning Up

These days, spillages and explosions at oil rigs are very widely reported—sometimes accurately. In terms of loss of life, the worst oil disaster was the explosion on board Occidental Petroleum's Piper Alpha platform in the North Sea in July 1988, when 166 oil workers were killed. Despite Piper Alpha, few people in the Western world are killed in a typical year in the oil industry compared with other energy industries such as coal.

The financial and environmental costs of oil spills seem to generate more column inches of comment than the cost in human life. The public is increasingly concerned with environmental degradation, and in the West this concern matters to politicians, who can and do extract large financial penalties from oil companies that spill oil: the much-publicized Deepwater Horizon spill in the Gulf of Mexico may cost BP $20 billion in cleanup and litigation costs. I summarize this most recent disaster later in this section; here, I want to put it in context by considering the general trend of oil spills over the years. I am not going to argue one way or the other about the ecological consequences of oil spills; it seems to me that nobody really knows, and in any case, environmental impacts are not the subject of this book. The economic cost of oil spills *is* part of this book because it is influencing the debate about the future of oil and causing oil companies to change their game plans.

In figure 5.5 I have put together the available data on reported oil spills. These data, obtained from Internet websites such as Wikipedia, are all supported by credible records. The spillages shown for each year are minimums, for several reasons. First, it is sometimes difficult to estimate the

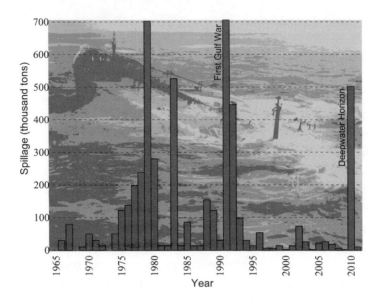

FIG. 5.5. Total oil spillage per year, 1966–2011. These are minimum numbers for each year, for reasons outlined in the text. The oil from some spills lingered for several years; for each of these I have simply divided the total spillage by the number of years, and added this figure to the total for each of the years of spillage. *Compiled from data gathered from Internet sources (such as Wikipedia's list of oil spills) where credible source references are provided.*

number of tons of oil that leak out from a well after an accident, so a range of values is often quoted in the literature. For example, sources indicate that the amount of oil spilled when the *Amoco Cadiz* went down off the coast of Brittany in 1978 was 223,000–227,000 tons (an unusually precise figure.) I adopt the lower figure. Second, for some spills the volume of oil lost was not recorded. Third, not all spills will have been reported.

Note that the graph shows no clear trend, except that some years are particularly bad, for reasons that vary. Not shown in the graph are the *number* of spillages for a given year, typically 1–10; again, some years are especially bad (e.g., 2005, as a consequence of Hurricane Katrina, though the total spillage for that year was unexceptional).[33]

The worst spillage in the history of oil extraction is not shown on the graph. It occurred between March 14, 1910, and September 10, 1911, at the Lakeview gusher in California. No less than 1,230,000 tons of crude spewed out of the ground, creating a lake of oil that covered 60 acres. This spill is almost twice as large as the total for the second-worst year (either 1979 or

1991). Cleanup was not an issue in those days, but today it is a huge undertaking.

DEEPWATER HORIZON

The largest accidental spill in American history, and the most financially costly, was from the Deepwater Horizon. Millions of words have been spilled (along with the oil), so I will restrict this account to a bare-bones summary and a brief consideration of the consequences of this disaster for the energy industry. The spill resulted from a methane gas explosion on April 20, 2010, that killed 11 oil workers on the Deepwater Horizon rig at the Macondo Prospect oil field in the Gulf of Mexico, some 40 miles southeast of the Louisiana coast. This rig was owned and operated by Transocean, the world's largest offshore drilling contractor. Transocean had leased the rig to BP and was operating it for BP. The explosion caused a fire which burned unabated for two days, until the weakened rig collapsed into the sea. The pipeline was severed at the seabed, a mile below the surface, and spilled oil into the sea. The depth of the water made it very difficult to cap the wellhead, and so crude oil spewed into the Gulf of Mexico for three months. A total of 4.9 million barrels of oil leaked into the sea, much of it rising to the surface and washing up on the local coastline and into coastal wetlands (fig. 5.6).

Finger-pointing and litigation, along with cleanup, began before the wellhead was capped. BP took most of the blame and consequently was sent most of the bills—some $20 billion at the time of writing. Local businesses such as fishing and tourism were degraded along with the environment. In May 2010 President Obama issued a moratorium on new drilling in the Gulf of Mexico, citing irresponsible action on the part of all the large oil companies that were active in the area. Some environmentalists argued that the ban did not go far enough; business argued that it was detrimental to local industries; political opponents of the president made hay while the oil leaked. After six months the moratorium was lifted.

The uneven playing field between the oil business and average citizens came under heightened public scrutiny as the media delved into the murk. Although there were already environmental measures that set the ground rules for oil exploration in the Gulf of Mexico, large oil companies operated under environmental waivers, so these measures did not fully apply to them. Even during the moratorium, drilling permits with environmental waivers were issued. Meanwhile, the smaller "coyote" companies were suffering because they lacked the resources to ride out the lean months following the disaster.

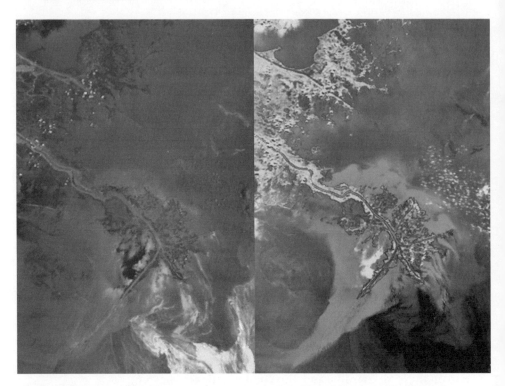

FIG. 5.6. Two satellite images of the Gulf of Mexico showing the BP Deepwater Horizon oil spill. Oil is inundating Louisiana coastal wetlands. The photos were taken using different parts of the electromagnetic spectrum. *Image from NASA/GSFC/LaRC/ JPL/MISR team.*

There was of course an investigation of the disaster. The report of the National Commission on the BP Deepwater Horizon Oil Spill and Offshore Drilling, when it was released in January 2011, made sweeping recommendations covering offshore safety, environmental safeguards, spill response measures, well containment mechanisms, and financial responsibility. Political and economic consequences spread much wider than the oil slick: BP is the largest corporation in the United Kingdom, and the disaster affected the British economy and pension funds. Such is the size of BP, however, that even with a $20 billion hit, it was well on its way to recovery by March 2012.[34]

The oil industry has taken note of BP's fate and will adapt, of course. No doubt the insurance cost for operating rigs has increased in the post-disaster environment. If it remains economical to recover oil from beneath the Gulf of Mexico, the rigs will continue to operate, and new ones will be towed out to newly discovered fields there. If not, the wolf pack will move on to find happier hunting grounds, such as those new hot spots off the Angolan and Brazilian coasts.

TRANSPORTATION ACCIDENTS: VULNERABILITY, AGAIN

Apart from oil rig explosions and refinery explosions, thousands of documented explosions, fires, and leaks have resulted from accidents that occurred during the transporting of oil and gas. The *Exxon Valdez* tanker ran into a reef in Prince William Sound in Alaska in 1989, spilling over a quarter million barrels of oil (perhaps three times that amount) and generating a massive and well-publicized cleanup. The litigation may last longer than the environmental consequences. In the same year, two passing locomotives emitted sparks that ignited gas from a leaking pipeline near the city of Ufa in the Soviet Union, killing 550 passengers, mostly children. In 1998 an oil pipeline exploded outside Lagos, Nigeria, resulting in hundreds of deaths.

The long list of such disasters serves as a sad reminder of the vulnerability of oil and gas infrastructure to accidents, as well as to intentional damage. In developed countries the loss of life tends to be lower, but it is less clear that spillages are also lower (data tend to be skimpy for spillages in undeveloped or developing regions). It seems likely that these accidents are as inevitable as car accidents—and that we will accept them (while trying to minimize them) as the price we pay for our automated lifestyle.

Consumption Resumption

The chart in figure 5.7 combines data on population, area, and oil consumption for seven developed nations.[35] These data show a clear inverse correlation between per capita oil consumption and population density (indicated by numbers at the top of the bars). This is not at all surprising: we expect longer journeys in large countries and therefore the consumption of more transport fuel. Note the exception, however: the United States lies in the middle group of population density but at the top end of the consumption chart. (Australia and Canada have much lower population densities than Sweden, Finland, and the United States; Germany and Britain have much higher densities.) The GDP generated per barrel of oil consumed, shown by circles on the chart, falls as oil consumption rises. This also makes sense: a crowded country like Germany can do a lot of business locally, whereas

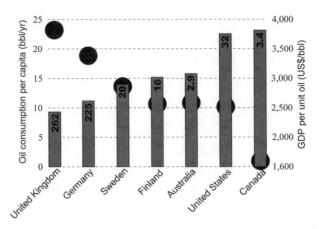

FIG. 5.7. Per capita oil consumption (*left vertical scale*) and GDP per unit of oil consumed (*right vertical scale*) for seven developed nations. Oil consumption is shown by bars; the number at the top of each bar is population density (number of people per square kilometer of land); the GDP per barrel is shown by circles. For comparison, world per capita consumption is about 5 barrels per annum—and it is less than that in rapidly modernizing China and India. *Data from* CIA World Factbook.

more mileage generally has to be covered to do the same amount of business in Canada. The U.S. entry looks a little low for the country's population density. Conclusion: the large amount of petroleum that each American consumes in one form or another per year has little to do with business activity and less to do with distances traveled. The suggestion is that it may be possible for America to cut its oil use without adversely impacting the lifestyle of its citizens. Specifically, the United States would be in line with the other affluent countries of figure 5.7 if it cut consumption by 10%–20% without adversely affecting GDP per capita.

The oil industry processing chain, from crude in the ground to gas in the tank, begins with upstream work—exploration and drilling—both of which are very expensive and hi-tech. Extraction of oil proceeds in phases and becomes increasingly difficult and expensive as a well is depleted. Finally, the downstream work consists of refining and transporting the petroleum products—mostly gasoline and diesel and aviation fuel. The entire processing chain is vulnerable to accidents and sabotage. Since oil is a fossil fuel, its production and consumption cause significant pollution and environmental damage. Oil reserves are being depleted, and oil consumption will probably exceed production within the next 10 years. Thereafter, the scramble for oil will depend more and more upon unconventional sources of synthetic crude, such as shale and tar sands. Eventually, we will need a replacement for oil.

6

WATER, WATER, EVERYWHERE

We move on from fossil fuels and consider our first renewable source of energy: hydroelectric power. Hydro is by far the best-developed, most mature technology for generating electricity from a renewable resource, and this fact is reflected in the figures: worldwide, hydro is responsible for 99% of the renewable energy we exploit.

Globally, installed hydro capacity is currently something like 890 GW, which is about 6% of the total of all available power. It is used mostly for electricity production, of which it constitutes 16% of the total. One of hydropower's advantages is that it can be turned on and off quickly and relatively easily; it can thus cope with the fluctuations of demand that we explored in chapter 1. In fact, those countries not blessed with a geography that permits much hydropower utilize what little they have almost entirely for peak-load demand. We can tell this from the figures: 3,400 TWh of electrical energy per annum and 890 GW of electrical power amounts to only a 44% *capacity factor*—the fraction of capacity that is in use at any given instant, on average. Hydropower depends upon falling water, as we saw in chapters 1 and 2, and so its adoption varies considerably around the world. Almost all Norwegian power—the base load as well as the peak load—comes from hydro. Fifty-eight percent of Canadian power is hydro, but it accounts for only 7% of American power. Yet, within the United States, dependence on hydro varies widely from state to state: Idaho, Oregon, and Washington rely upon hydro for over 80% of their power.[1]

The use of hydro power has more than doubled since the mid-1960s; recently, this increase has been particularly rapid in Asia. Studies undertaken globally provide interesting figures for the future, untapped hydropower capability of each of the world's major regions. In developed regions like North America and Europe there is relatively little untapped capacity because suitable sites for hydroelectric dams are either already in use, are protected for environmental reasons, or are reserved for other purposes. Africa has the most potential; less than 10% of its hydro capacity has been tapped. Over the next 20 years, however, the regions that will most rapidly develop their hydroelectric generating capacities are Asia and Latin America. China is pushing forward its development in this key technology (indeed, in most technologies) very aggressively. Currently, 17% of China's power comes from hydro; the hydro capacity of that nation is due to increase by over a third—an increase of between 49 GW and 65 GW capacity.

Worldwide, the economically feasible hydro potential is around 8.77 million GWh per year. This number represents an increased hydro capacity worldwide of 1 TW power, assuming 100% utilization. This sounds terrific until we realize that a capacity of 1 TW power is only 5%–7% of current worldwide demand. Hydropower is not going to be the silver bullet that we need. It is and will remain an important power source—and a relatively benign one, as we will see—but it is not going to be our main source of power in the future.

Of course, for some countries hydro will be a highly significant source of power. In Canada, for example, much of the untapped hydro potential is earmarked for the U.S. market. The planned capacity and economically feasible capacity in Canada would more than double the country's already large hydro capability; thus, hydropower could provide more than 100% of Canada's power, with the excess being sold to its southern neighbor.

Compared with other power sources, hydro power is inexpensive. The unit cost of a gas turbine power plant is about 3.7 cents/kWh of output energy. Fossil fuels cost 2.2 cents/kWh; nuclear power, 1.8 cents/kWh. Hydro is 0.9 cents/kWh. Hydro is inexpensive because its fuel (river water) is free and because a hydropower plant (a dam plus turbines) is robust and lasts a long time. On the other hand, the initial investment is very high, in both cost and time. The capital cost per kilowatt-hour for the world's biggest hydro project, the Three Gorges Dam in China, was $1,200. For the Brazilian/Paraguayan Itaipú plant, the world's second largest hydro plant in terms of output power, the capital cost was $1,600/kWh. Labor costs for both these projects were much lower than they would have been in the developed world; an American hydropower project costs $2,000–$4,000/kWh. Thus, hydro has its faults as well as its virtues, but in the hardheaded world of today, the bottom line looks good for hydroelectric power.[2]

Scaling the Heights

The construction cost, power generated, efficiency, and environmental impact of a hydro plant vary with scale in ways that are not simply proportional to the size of the plant. Thus, for example, a million tiny plants each producing 5 kW of power will have a much smaller environmental impact than a single giant plant producing the same total power, 5 GW. This is because a tiny hydro plant (sometimes termed *pico hydro*) is probably a simple waterwheel in a backyard, supplying power for a pump or for domestic lighting for a single house. The waterwheel does not cause any flooding or build up any head of water beyond the boundaries of a seasonal stream or creek. A million such units have a million times the environmental impact: a million zeros is zero. Contrast this with the substantial impact of a giant dam, of which more shortly.

Because the character of hydro plants depends upon their size, hydro power is often divided by scale. People talk about pico hydro, micro hydro, and so on, though there is no universally agreed range for each of these scales. Consequently, here I will make up my own names and ranges. The main point I want to convey is the diversity of sizes; what descriptors are applied to the plants and what their exact sizes may be are not really important. Let us divide hydro plants into micro, mini, medium, mega, and mammoth. A hydroelectric power source is here classified as *micro* if it generates less than 100 kW of power. Such a plant would serve a single house or a few houses or a small family business. A lot of micro hydro plants exist in the third world.[3] I will call a hydro site *mini* if it produces more power than a micro site but less than 15 MW. Such a plant might be constructed for a community or for an industrial facility. A *medium* hydro plant generates up to 100 MW, while a *mega* plant produces up to 1 GW. Anything above 1 GW is *mammoth*. These last three plant types are usually constructed to provide electricity for a large energy grid.

One aspect of hydropower generation that changes with scale is the plant *efficiency*. Plant efficiency is the ratio of actual power generated to available water power. Efficiency increases with scale, from 40%–50% for micro plants to over 90% for mega and mammoth plants. There are differences in turbine technology as we change scale, and the details of plant efficiency depend upon turbine type. However, we can understand the general trend (increasing efficiency with increasing size) from simpler and more general geometrical arguments. We saw in chapter 2 how to estimate the power that can be generated by a plant: it depends upon the water flow rate and the height of the dam (water *head*). The flow of water is a volume effect. Its increase with scale is proportional to the volume increase of water that flows through the plant in a given time.

On the other hand, power is wasted through frictional effects, which are surface phenomena. Thus, turbine parts rubbing against each other sap power, the power loss depending on the area of contact. Water flowing through *penstocks*—the pipes that carry headwater down to the turbines in hydroelectric plants—loses power as a result of friction between water and pipe surface. The point is that, as the size of a plant increases, both surface area and volume increase, but volume increases faster. To express the concept mathematically, if we scale up a dam that stores water for generating hydro power, the frictional effects will increase as the square of dam height whereas the power generated will increase as the cube of dam height. Thus, frictional loss becomes relatively less important as scale increases—bigger is more efficient.

Another advantage of large plants is the operating and maintenance costs per megawatt of power produced. Unsurprisingly, it is less expensive to produce a megawatt of power from a large hydro plant than from a small one. A medium hydro facility operates at perhaps a quarter of the cost, per unit power produced, of a mini hydro site, for example.[4] Equally unsurprisingly, maintaining 10,000 micro hydro plants costs more than maintaining a single mega plant producing the same power (think of the travel costs of maintenance crews).

Per megawatt of power produced, small plants win out in initial construction costs. For example, small sites do not require long and expensive geological surveys, and they do not produce large headwaters (*pondings*) that displace people and destroy habitat (more accurately—and this nicety is significant—visibly destroy habitat). Small sites are quick to build and do not require long transmission lines. This last point brings us to an aspect of hydropower—location of the dam—that merits a separate paragraph.

Small dams, say for a micro or a mini hydro facility, can be built almost anywhere there is running water—across a creek or stream or in a small river. Medium dams require larger and therefore less numerous rivers, which often are already being used for other purposes (inland ports, transport, fishing, irrigation, etc). In addition to large water flow, mega and mammoth dams require suitable geology. Very large dams are very heavy and can sink into the ground unless the underlying ground is sufficiently robust. Not only geology

---

PEOPLE POWER

Assuming that the average person consumes 3 MWh of energy each year, and assuming a capacity factor of 40% for hydropower, we can estimate the number of people served by, for example, the Hoover Dam hydro plant.

Three megawatt-hours of energy converts into an average per capita power consumption rate of about 350 W.* Hoover Dam provides 2 GW peak capacity, or an average of 800 MW, given a capacity factor of 0.4. This means that the Hoover hydro plant can supply the power needs of about 2.3 million people (at 350 W per person). Currently, Nevada receives 23.37% of its power from Hoover, corresponding to 530,000 Nevadans (one-fifth of the state). Los Angeles receives 15.42%, and so 350,000 Angelenos should receive their power from Hoover. Boulder City—created near Hoover Dam to house its construction workers—gets 1.77% of the power, which is enough for 40,000 people (good news for a city of 5,000).

Except that Americans are not average people when it comes to energy and power consumption. In fact, the average American consumes power at the rate of 1,360 W, which is 3.9 times the world average. Thus, only 590,000 Americans get their power from Hoover Dam—135,000 Nevadans, 90,000 Angelenos. When Boulder City doubles its increasing population, it will have to ask for a bigger slice of the Hoover pie.

*In figure 2.1 we saw an average power consumption for modern humans of over 11 kW. However, that figure included power generated by industry, street lighting, and other nondomestic uses, whereas here we are considering in-home power use only.

---

but also geography needs to cooperate for the large facilities. Deep or wide valleys with strong and impermeable walls are required. The combination of requirements means that there are only a limited number of sites around the world that are suitable for very large hydroelectric dams. If such a site is close to population centers, then massive population displacement occurs before the dam headwaters flood villages and towns (1.24 million people were relocated to make way for the Three Gorges Dam, for example). If the site is not close to a population center (e.g., Hoover Dam), then there are significant costs in transmitting the electrical power that is generated to where it is needed.

Valleys of the Dammed

Mega and mammoth hydro dams, as we just saw, have to be located in valleys. The geology and geography of the valley determine the type as well as the size of these larger facilities. Such dams come in four basic types:

- embankment dams,
- gravity dams,
- buttress dams, and
- arch dams.

Embankment dams are often very large but are not correspondingly high. That is, embankment dams have shallow, sloping faces with very wide bases. They are made from earth fill or rock fill, both obtained locally and therefore much less expensive than concrete. Because of the wide base, ground pressure is not a problem, even for the largest embankment. Consequently, unlike the other large dam types, embankment dams can be built on open-jointed (loose or porous) rock. Because an embankment dam is essentially a large pile of dirt placed across a river, such dams are not very susceptible to earthquake damage. They do leak, however, and this is the weakest aspect of this design type. Since the basic building material is not watertight, embankment dams are faced with clay, dressed stone, or (nowadays) concrete. The Aswan High Dam in Egypt (with a 2.1-GW capacity—a mammoth) is of the embankment type.

Gravity dams are made of concrete. Whereas an embankment dam generates hydropower by virtue of its width (and hence, large flow rate), a gravity dam generates the same power by virtue of its height. Dense concrete cannot be easily pushed aside by water; thus, gravity dams can be quite high. They are expensive to build; not only is a huge amount of concrete required, but efficiency and costs often dictate the construction of a special factory near the dam site to make that concrete. The underlying bedrock needs to be strong enough to hold the weight, and gravity dams are consequently not as wide as embankment dams. Gravity dams stretch across narrow valleys with nonporous walls, allowing headwater to build up quite high pressure in front of the dam. These dams are usually straight and of simple design. The Chief Joseph Dam on the Columbia River in Washington State is a gravity dam with a mammoth hydropower capacity.

Buttress dams are reinforced concrete walls and are thin compared with

FIG. 6.1. A buttress dam in France. Image by Versgui.

gravity dams. They are buttressed on the downstream side, to resist water pressure (fig. 6.1). While they are less strong than gravity dams, they are also much less bulky and in many ways are a compromise, with their advantages and disadvantages both ameliorated when compared with gravity dams, from which they evolved. One aspect of buttress dam design that is superior: they are more versatile, in that the design can be more complex or variable in adapting to local conditions.

Of the large dams, arch dams are the most visually elegant. Like buttress dams, they are thin concrete walls, but instead of being buttressed, they are curved like an arch on its side. The ends of the dam are braced against strong valley walls. The huge forces acting on the dam from water pressure are transferred along the dam to the supporting rock walls. Unlike gravity or buttress dams, arch dams depend on the compression strength of concrete, rather than its weight, to hold back the water. Arch dams are not heavy, requiring relatively little concrete. Their main disadvantage is the strict requirement for narrow gorges or valleys with strong, impermeable rock walls; there is no point in bracing an arch dam against fractured rock or earth that is going to yield as soon as the water pressure builds up against the dam.

Combinations of these basic dam types are common. Two of the largest of these mongrel designs are the Daniel-Johnson Dam in Quebec, which is a multiple-arch-buttress dam, and Hoover Dam in Nevada and Arizona, which is an arch-gravity dam (see fig. 6.2). Hoover Dam is arched, but the lower portion is very thick and heavy. Both of these dams are mammoths in my nomenclature: they each have a capacity exceeding 2 GW.

These four basic designs apply to any type of dam, not just to dams designed as part of a hydroelectric plant. They are called *impoundment dams* because they impound (store) river water. There is another type of dam designed solely for hydro purposes, however, and that is the run-of-the-river dam. These do not store water permanently and, indeed, have little pondage: the running river water is directed through penstocks to the turbines. Such hydro facilities are subject to seasonal variations in water flow rate, but on the other hand, they do not disturb the river environment by creating large headwaters. The normal course of a river that is dammed in this way is not altered. These dams are usually of medium size or smaller, but one mammoth example is the Chief Joseph Dam.[5]

Turbines

Apart from waterwheels at the smallest, micro scale, hydropower is generated by turbines. (Some engineers classify waterwheels as turbines, denoting them *cross-flow* turbines.) The optimal turbine design for any given facility depends upon flow rate and water head—in other words, upon scale. The largest plants, including the Grand Coulee on the Columbia River in Washington State, use Francis turbines (see fig. 2.7). The Grand Coulee is a mammoth dam with a capacity of over 2 GW. The world's largest dam, the Three Gorges in China, has 10 times this capacity, and it too makes use of Francis turbines.

Between the extremes of cross-flow and Francis turbines there are a host of other designs. They differ in subtle nuances of blade shape, a detail that does not concern us here. Suffice it to say that each turbine finds its niche at a particular flow rate or for a particular head of water. Thus, the Kaplan turbine, a propeller type with adjustable blades, is most suitable for hydro facilities with low hydraulic head (2–40 m). The Turgo turbine is designed for medium head (50–250 m) and attains an operating efficiency of 85%–90%. The Pelton turbine works efficiently when the hydraulic head is large and flow rates are small. All of these designs date from either the nineteenth

FIG. 6.2. Hoover Dam. (a) The massive bulk of this arch-gravity dam can be seen braced against the solid rock walls of Black Canyon, on the Arizona-Nevada border. At 726 feet high, Hoover Dam was the world's largest when it was completed, during the Great Depression. (b) The dam created Lake Mead, covering 247 square miles. (c) The hydro facility at the base of the dam efficiently converts the potential energy of water into gigawatts of electricity. *Photos by the author.*

century or early twentieth century; water turbines have been an important part of electricity production for a long time.

Technically, turbines fall into one of two categories depending upon the physics of their operation. *Impulse turbines* work because flowing water transfers its momentum to the turbine blades, causing rotation. Cross-flow, Pelton, and Turgo turbines fall into this category. *Reaction turbines* work because the blades react against the flow of water, moving in the opposite direction, like a garden sprinkler. Francis, Kaplan, and propeller turbines are of this type. Large water turbines are wonderfully efficient (around 90%) when operating in their comfort zones, and this efficiency (plus, of course, the clean and inexpensive fuel) is a significant reason that hydropower is seen as environmentally benign.

Risks and Rewards

Now it is time to examine the advantages and disadvantages of hydropower, so that we can fit it into the scheme of things as we consider future global power sources.

These advantages and disadvantages are readily listed. In the following text I will add flesh to the bare bones.

Pluses
- Hydropower is clean; no toxic materials are used to generate it, and no toxic materials are produced from it.
- It is domestically sourced (usually; there are notable exceptions) and so is not subject to the winds of international politics or economics.
- It is renewable in the sense that rain will always fall, though the rate may be variable.
- As we saw in chapter 1, it enables a power generating station to act like GoD (*G*enerate *o*n *D*emand). By controlling water levels, hydropower stations can control output power.
- It provides auxiliary benefits, including increased irrigation from the headwaters, increased navigability of the upstream river system, recreational use of the headwaters, and increased fishing.

Minuses
- The initial construction costs of large dams are *huge*, and maintenance costs due to siltation of the headwaters can be high.

- The specialized geology and geography required for big dams greatly restricts possible sites. If nature requires the site to be in the middle of a remote jungle, the environmental cost of dam construction can be great, and electricity transmission may be expensive.
- The enormous scale of large dam construction and the huge headwater area can mean considerable population displacement and social disruption.
- Downstream of the dam, water flow is more sluggish and carries less silt than it did before the dam was constructed. These facts can have significant negative effects many miles downstream, in the form of reduced dissolved oxygen and fewer nutrients.

A little more detail will shed light.

Hydropower is clean. Yes, it is: the fuel is water, which is environmentally cleaner than oil. It is not quite true to say that no toxic materials are produced by hydropower, however. Initially, as the headwaters are forming behind a dam, some methane (a greenhouse gas) is produced by decaying vegetation that suddenly finds itself underwater. For impoundment dams, the area of land thus affected can be very large, as we have seen. However, this is a one-time cost, and it is significant only where a lot of vegetation existed in the pondage area. (Thus, Lake Mead is huge, but the area covered by water was desert, with only a low density of vegetation.)

It is domestically sourced. This is an important factor and will arguably become more important as oil and coal sources become more and more scarce. One benefit of such secure sourcing includes price stability—so vital to future development, as anybody in the oil industry will attest. More obviously, the energy is more secure. Even if some local nefarious ne'er-do-well wanted, for reasons best known to himself, to destroy a hydropower plant, he would find it very difficult: a shoeful of semtex will not do much to six million tons of concrete (approximately the weight of Hoover Dam). Also, hydropower sites, unlike pipelines, are relatively small and easily protected.

It is renewable. So long as rain falls, and the river that forms from runoff is not diverted for other reasons (such as irrigation), then the fuel source is sustainable. Gravity is free (indeed, obligatory)—so the energy of falling water is there to be tapped for as long as the water flows.

It can be generated on demand. Hydropower is often used to satisfy peak demand because it can readily be turned on and off. The rotation direction of the water turbines can be changed in just a few minutes, so that instead of generating electricity, the turbines can use excess electrical power to pump water uphill and store it until needed.

It has auxiliary benefits. The headwaters that are created to increase the height of water falling onto the turbines provide secondary benefits up- stream of the dam. Farmers have access to irrigation water that was pre- viously too low for them to use without expensive pumping. The navigable stretch of river upstream of a dam is increased, in some cases for hundreds of miles. This can be an important economic factor, reducing local transporta- tion costs. The controllable water level means that navigability is less prone to seasonable variations in water levels. Fishing can be improved (because of the increased habitat for fish) or made worse (for instance, when dams prevent salmon from returning upstream to breed, as at the Chief Joseph Dam).

Hydropower incurs costs. Turning to the disadvantages of hydropower, a politically obvious one is the high cost of construction. A lot of work and money must be expended before any benefit is obtained—though this is true for any power source. Because dammed rivers flow more slowly than they did before being dammed, they tend to silt up; siltation, if left unchecked, will choke the dam. (Particulate matter in the water also wears out the turbine blades.) It is expensive to maintain silt-free headwaters, but despite this drawback, hydropower running costs remain competitive, as we have seen.

The sites suitable for dams are limited. Here we have the crunch issue, in a nutshell. The geologic and geographic requirements for dams, at least for the large-scale (medium to mammoth) impoundment dams that provide most of our hydropower, severely limit the number that can be created. (And as more and more remote hydro sites are exploited, the low cost of hydropower will increase.) The world has been mapped for its as-yet-untapped hydro- power potential; the results show that hydropower will not be able to supply more than a small fraction of the world's energy needs. This is especially true in (most) developed countries, where easily accessible or readily dammed hydro sites already have dams. The uneven distribution of hydro sites across the globe adds a political twist to future developments, perhaps compromising the domestic-source advantage currently enjoyed by hydro. Will African nations in 50 years' time be selling hydropower to Europe? Will Canada extract trade concessions or higher charges from its energy-addicted neighbor to the south? Will hydropower-rich regions such as Norway diverge in energy policy from friendly but hydropower-poor neighbors such as Denmark?

Dams cause disruption. There was a considerable criticism of Chinese authorities over the decision to move 1.24 million people from the 400 square miles of valley that later held the headwaters of the immense Three Gorges Dam: the people were given no choice in the matter. Imagine the litigation that would have resulted from such action in the Western world. If we consider that most oil and gas sites are remote, that most coal mines are remote, and that nuclear power plants are small, we can see that more population disruption occurs from impoundment dam hydropower development than from other energy sources. (Wind power takes a lot of surface area, but it is mostly offshore or on sparsely populated hilltops. Although the low energy density of solar power plants means that they spread over a large area, most sites suitable for large-scale development are in sparsely populated deserts.) Population disruption and social dislocation are factors that will not affect all countries equally: thus, China, with its dense population even in the mountainous southeast, will feel the effects far more than Canada, whose hydro sites are remote from population centers.

Dams can have adverse downstream effects. While upstream regions gain navigability and perhaps fishing from the construction of dams, downstream regions lose navigability and fishing. Because the dam reduces the flow of silt downstream, flood plains are less fertile. Dams also make the downstream water more sluggish, which means that it carries less oxygen as well as fewer nutrients. Consequently, the water itself as well as the riverbank area is less productive. Reduced flow rate also influences riverbank erosion patterns. Such effects are measurable for hundreds of miles downstream of a large dam. Another, less obvious downstream effect of a dam, a mirror image to the increased irrigation prospects upstream: reduced aquifers and hence reduced well water and irrigation.

Thus, while the environmental effects of hydropower are mild compared with the effects of other major power sources, they do exist.[6]

Seapower

Hydropower does not have to be restricted to freshwater. Indeed, from a purely hydrodynamic perspective seawater should be 2.5% better than freshwater, being denser by that amount. Yet relatively little of our hydropower comes from the sea. One blindingly obvious reason for this is the lack of hydraulic head—the sea tends to be at sea level. Nevertheless, the movement of seawater provides two possibilities for hydropower: tidal power and wave power.

TIDAL POWER

Tides arise from the gravitational interaction of the earth with the moon and sun. The power that moves the world's oceans from place to place, twice a day, is dissipated as friction.[7] This friction causes a slow but sure depletion of gravitational potential energy, resulting in the moon's moving further away from the earth and the earth's rotation slowing down very gradually.

The tidal forces are huge, and so surely, you would think, it would be possible for mankind to tap into these forces and generate power from them. Yet the difficulties are severe enough for tidal power to be relatively unimportant, even though its potential has been recognized for centuries. Research over the last few decades has renewed interest in the subject, and some sizeable tidal power plants will appear within the next few years.

Tides are fairly predictable and reliable, but there are few sites around the world where it is feasible to place a power plant that taps into tidal power. The topography needs to be such that tidal water is channeled into small areas so that the difference in height between high and low tide is appreciable (5 m, or 16 feet, is the working rule of thumb for tidal power applications). At the same time, the underlying seabed must be suitable for the placement of a stable power generation facility. There are two types of tidal power stations. *Tidal barrage* power plants take advantage of the hydraulic head difference between high and low tide levels (i.e., they depend on the potential energy of the water), whereas *tidal stream* power plants take advantage of the water flow (its kinetic energy). For a tidal barrage, the seabed must support dams that are strong enough to retain ebb water[8] and to avoid erosion due to flow in two directions. In a tidal stream plant, turbines are affixed to the seabed and must be placed in regions of strong currents that flow first one way and then the other.

The way in which tidal barrages extract energy is shown conceptually in figure 6.3a. Quite a few medium-sized tidal barrage power stations exist, but there are only two mega plants: the Rance tidal power station off the coast of Brittany, in northwestern France, which was built in 1966 and has a capacity of 240 MW; and the Sihwa Lake station in India, completed in 2011 and with a capacity of 254 MW. A mammoth site—the Incheon facility off the coast of South Korea, which will have a capacity of 1.3 GW—is under construction and due to be completed in 2017. Much larger facilities are on the drawing board—and may or may not stay there, such are the technical challenges. Thus, for example, there are gigawatt-capacity facilities under consideration for Penzhin Bay, off the coast of the Russian Far East. The Severn barrage, spanning a large tidal inlet on the southwestern coast of England, has been proposed and analyzed several times, and rejected as many times. The world's highest tides are probably those at Minas Basin in the upper reaches of the Bay of Fundy, off Nova Scotia in eastern Canada. One hundred billion tons of seawater flow in and out of the Bay of Fundy with each tide—more than the total of all freshwater river flow. Surely an ideal place for a tidal barrage? There are research facilities but no large-scale commercial tidal barrages.

The problems of tidal barrages are financial as much as they are technical, and environmental as much as financial. The high capital investment required (typical for mammoth power plants of all types, but in this case for plants of untried technical reliability—each site is different) has sunk many a promising tidal barrage project while still on the drawing board. Tidal barrages also cause large and highly visible environmental changes—saltwater wetlands are damaged, shoreline ecology is changed—in regions where people like to visit or live. Littoral regions of the world contain the highest density of biomass because of the happy combination of warm waters, sunlight, and oxygen; but these very regions are the ones where tidal barrages must be located, so an environmental cost is part and parcel of this type of tidal power.[9]

The second type of tidal power plant, the tidal stream power station, has a much smaller environmental footprint but suffers from the fact that it does not exist—not yet, anyway, in any significant way. Tidal stream turbines can be thought of as underwater wind turbines or as run-of-the-river hydropower plants on steroids. These turbines turn because of the tidal currents flowing past the blades.[10] They have relatively small environmental effects and are largely out of sight (a significant factor in forming public opinion, sadly perhaps). Small tidal stream power plants exist in the Bay of Fundy and elsewhere—for research and development purposes—but the technological

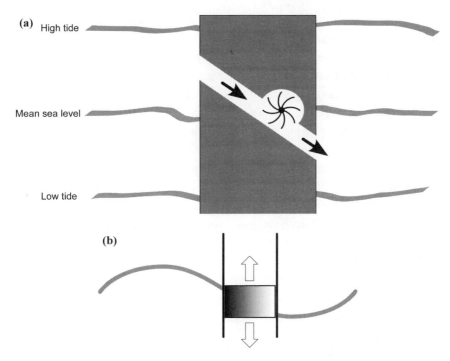

FIG. 6.3. Power from seawater. (a) Tidal barrage basics. Here the tide flows in from right to left. When it recedes, water flows down the penstock (arrows) causing a turbine to turn. A basic design choice is evident in this simple diagram: how high up, and at what slope, do we place the penstock? A little thought will convince you that the choice is a trade-off between generating capacity and generation time, between tides. Stated differently, the extracted energy is limited. (b) Wave power basics. In this simple schematic, a floating piston head is constrained to move vertically. Wave motion thus can be used to drive a piston.

challenges mean that this form of tidal power is some decades away from being commercially feasible and significant. As with all hydropower, even when fully developed it cannot become a major player on a global scale because of the limited area of suitable sites.[11]

WAVE POWER

Another form of hydropower derived from the sea—extracting power from waves—is quite independent of tidal power. Ocean waves originate from solar power (via wind power, which in turn originates from differential heating of the earth's surface) and not from gravity. One issue wave power shares with tidal power, however, is the technical difficulty of making power generation commercially feasible. When I was a graduate student at Edinburgh University, Scotland, in the late 1970s, one of the engineering faculty members was Stephen Salter, who developed the *Salter duck*, a device for extracting energy from ocean waves. It was hailed as a promising alternative for our dependence on oil during that period—the immediate aftermath of the 1973 oil embargo—and yet since then, we have heard little of it or of tidal power in general. A simple conceptual indicator of how power can be extracted from waves is shown in figure 6.3b.

The world's first commercial wave farm was the 2.25-MW facility at Aguçadoura, off the northern coast of Portugal, which opened in 2008. A second phase of construction would have increased the capacity by a factor of 10. However, the plant shut down within a few months of opening because of the global economic downturn of that year and technical difficulties with the equipment.

These examples suggest that tidal power is a technically difficult nut to crack and that large-scale operation is some way in our future. Even though a number of suitable sites have been identified—temperate zones off the Atlantic coast of Western Europe, off the Pacific coasts of North and South America, off South Africa, Australia, and New Zealand—as of yet these places have not been given over to wave farms. Should that occur, we can expect the same sort of environmental issues that arise for tidal barrage power plants. This is because wave farms reduce the amplitude of waves by extracting energy from them, just as tidal barrages reduce the height of tides, and because they will operate in the same littoral regions where much of life on earth resides.

Hydropower technology—for both dams and water turbines—depends on scale; it works more efficiently when hydro plants are large. Given that large impoundment dams require special geology and topography, the total hydro capacity of the world can never meet more than a small fraction of the global power need. Building dams for hydropower does have environmental consequences, but these are minor compared with other power generation technologies, and some of the consequences are beneficial. Tidal power and wave power technologies are not well developed and will not become significant in the foreseeable future.

7

TOO CHEAP TO METER

Although nuclear power is technically not a renewable resource, it is often considered such because the fuel reserves are enough to last for much longer than the reserves of other power generation technologies—anywhere from 80 years at the low end of estimates to millions of years at the high end. The figure you hear depends on whom you talk to and upon what they include and exclude. In fact, both these extremes are valid in their own way, as I will explain in this chapter, which is about the technology, the potential, and the future of nuclear power generation.

Nuclear Physics 101

The purpose of this section is not to provide you with a Ph.D. in nuclear physics or to turn you into a nuclear power engineer. I will quickly introduce some nuclear physics and engineering fundamentals and then just as quickly move on. My aim here is, first, to show you that nuclear energy is fundamentally different from all the other kinds of energy that we know about and have tapped into over the centuries to serve our needs;[1] and second, to bring you up to speed on the basic physics so that you can better appreciate the technological issues that arise in the sections to follow.

STRONG FORCE

In our everyday lives we encounter two of the four fundamental forces of nature. We all have a working knowledge of gravity—the most down-to-earth of the forces, so to speak. In high school or university we learned a little or a lot about how Isaac Newton understood the workings of gravity to hold the planets in their orbits and to hold us to the earth's surface. The second force we regularly experience is electromagnetic. Magnets attract some metals, and they attract or repel each other. Static electricity causes hair to stand on end and causes bits of paper to stick to combs. More importantly, every single energy or force or power that you can think of and experience—gravitation excepted—is electromagnetic in origin. The chemical energy of fossil fuels is electromagnetic, as are the force of your muscles and the power of pneumatic drills. The tensile strength of steel wire and the pressure exerted by a vise are electromagnetic. In this section we will see why this is the case.

There are two other fundamental forces of nature, which we do not ordinarily experience, because they act over very short distances (much smaller than an atom) and are normally confined within the nuclei of atoms. The weak nuclear force and the strong nuclear force reveal themselves only when these nuclei split apart. It is the strong nuclear force that we will concentrate on, once the introductions have been made.

Figure 7.1 depicts two schematic representations of a hydrogen atom and two representations of a helium atom. Concentrate on the representations to the left. Hydrogen consists of one proton and one electron. Helium consists of two protons, two neutrons, and two electrons. These beasts—protons, neutrons, and electrons—are three animals from the elementary particle zoo.[2] All the elements in nature and all those man-made elements not found in nature are composed of these three types of particle, in different combinations. Thus, an atom of the element uranium usually consists of 92 protons, 143 neutrons, and 92 electrons. Note that the number of protons equals the number of electrons. Each proton has a positive electric charge, while each electron has a negative electric charge (of the same magnitude), and so every atom has no net charge—proton charge plus electron charge sums to zero. Neutrons are electrically neutral, so an atom of hydrogen or uranium or any other element can vary the number of neutrons. Normally, as in figure 7.1, hydrogen has no neutrons, but there exist in nature variants of hydrogen (*isotopes*) with one or two neutrons. Of interest to us, the element uranium has two significant isotopes: uranium-235, or ^{235}U, has 143 neutrons, while the much more abundant uranium-238 (^{238}U) has 146 neutrons. Both have 92 protons, by definition (if the number of protons were different, then the atom would not be a uranium atom).

Protons and neutrons are bound together in the atomic nucleus, a small dot in the center of the atom that contains almost all the mass (a proton and a neutron is each about 1,836 times heavier than an electron). The electrons orbit outside the nucleus, most of the time, and occupy a much larger volume. The laws of quantum mechanics determine the arrangement of electrons in their different orbits around the nucleus and the arrangement of

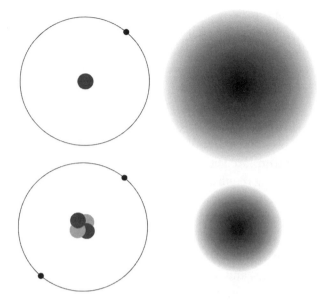

FIG. 7.1. Two schematic representations of a hydrogen atom (*top*) and a helium atom (*bottom*). Hydrogen normally consists of one electron (*small black circle*) orbiting a nucleus that consists of a single proton particle. Helium consists of two electrons orbiting a nucleus composed of two protons and (usually) two neutrons. Almost all the space occupied by an atom—any atom—is taken up by the electrons, though almost all the mass is contained in the nucleus. The renditions on the right side of the figure are more realistic, if less insightful (the fuzziness is due to quantum mechanical effects). Note that the helium atom is smaller than the hydrogen atom because its electrons are pulled in closer to the nucleus.

protons and neutrons inside the nucleus. All of chemistry involves only the electrons—it has nothing whatsoever to do with the nucleus. When sugar dissolves in water, or when sulfuric acid eats away at metal, or when iron rusts, paper burns, or glue sticks, the force at work is the force of electrons— the electromagnetic force. Nuclei tag along but are utterly unchanged by these electron shenanigans. Molecules form when electrons are shared between two or more atoms; again, these chemical reactions are between electrons and do not influence the nuclei or their proton/neutron constituents at all. In short, except for gravity, all of what happens in the everyday world is due to the electromagnetic interactions of electrons that orbit atomic nuclei.

And now for something completely different. When nuclei are made to interact—when the protons and neutrons come out to play, so to speak— then the strong force exerts itself. We are very aware of when this happens because the strong force is several hundred times stronger than the electromagnetic force and the energy released by the action of the strong force is so much more than the energy that we associate with electromagnetic actions. When we generate nuclear power, we are tickling the dragon's tail: inviting the strong force out into our world.

Here is how it works. If you weigh the nucleus of an atom—say the ^{4}He of figure 7.1—you will find that it weighs *less* than the sum of two protons and two neutrons. The whole is less than the sum of its parts. This is because the protons and neutrons are very tightly bound together by the strong force, and this tight binding requires a lot of energy. Where does this energy come from? Einstein taught us that mass is a condensed form of energy, that mass and energy are equivalent, and that one can be converted into the other. The rate of exchange is given by *that* formula—you know the one. The energy that binds protons and neutrons into a nucleus comes from the mass of the protons and neutrons, and it is called the *binding energy*.

The binding energy is not the same for all atomic nuclei. In figure 7.2 you

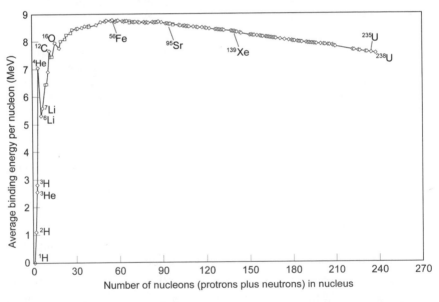

FIG. 7.2. Nuclear binding energy versus number of protons and neutrons in the nucleus. Iron-56 (^{56}Fe) is the most stable nucleus in the universe. Iron (symbol *Fe* for Latin *ferrum*) has 26 protons; the superscript 56 refers to the number of protons and neutrons in the nucleus of this iron isotope. Nuclear reactions that release energy must move along the binding energy curve toward ^{56}Fe; for example, ^{235}U splits into ^{139}Xe and ^{95}Sr when struck by a neutron. *Image from Wikipedia, s.v. "Nuclear binding energy."*

$$E = mc^2$$

Let us suppose that the current world consumption of power is 20 TW, a figure that is at the upper end of the various estimates that have been made (see the introduction). Let us further suppose that all of this power is generated by nuclear reactors. How much mass would need to be converted into energy each year to fuel our power needs?

From the power consumption figure we see that the annual energy consumption is 630,000,000,000,000,000,000 J. Using *that* equation—surely the most famous equation of mathematical physics—we divide by the speed of light (squared) and arrive at a figure of 7008 kg. That is, 7 tons of matter turned into energy each year would be enough to provide all our energy needs.

---

can see how binding energy varies with the number of nuclear particles (i.e., with the type of atom that a nucleus constitutes). An important consequence of the binding energy curve is that energy is released when nuclei with a relatively low binding energy can be persuaded to split into smaller nuclei with a larger binding energy. Nuclear bombs and nuclear power plants achieve this trick.

CHAIN REACTION

How is the nucleus of an atom split? A good method is to hit it hard with a neutron. Here is how a nucleus of the ^{235}U isotope reacts when it is hit by a neutron:

$$^{235}U + n \rightarrow {}^{95}Sr + {}^{139}Xe + 2n + \text{energy}.$$

Here, n stands for a neutron, and Sr and Xe are the elements strontium and xenon. In words: hit a nucleus of uranium-235 with a neutron, and the result is a nucleus of strontium-95 plus a nucleus of xenon-139 plus two neutrons plus a helluva lot of energy. Exactly how much energy can be found by examining the binding energy curve: it turns out to be about 0.000000000032 J. This may not look like much, but of course this is just one nucleus of ^{235}U. A kilogram of the stuff will split up to release 83 TJ of energy (23 GWh)— enough to power New York City for several hours.[3]

The important point to note about this nuclear reaction (apart from the enormous energy released) is that two neutrons are produced for every one that is consumed. These two neutrons can then interact with other ^{235}U nuclei and cause them to split, releasing more energy and more neutrons. This process is called a *chain reaction*. As I have described it, the number of reactions doubles at each stage of the chain, and so the amount of energy released doubles. Thus, the process rapidly explodes (literally) until all the available ^{235}U is used up. This is how a uranium bomb works. In a nuclear power plant that consumes ^{235}U fuel, the reaction rate is controlled by inserting into the reaction chamber material that absorb neutrons, so that only one neutron on average is released for every neutron that is absorbed. In this way the explosion is slowed down to a constant rate, sustainable for as long as ^{235}U fuel is fed into the reaction chamber. This is how one type of nuclear power plant works.

There are many other types of nuclear reactions, apart from the one I have described that splits ^{235}U. Other nuclei that are suitable for nuclear power plants include ^{239}Pu and ^{241}Pu, two isotopes of plutonium, and ^{232}Th, a thorium isotope.

FISSION AND FUSION

The nuclear reactions that I have described so far—ones that release energy by splitting a large nucleus into two or more smaller nuclei—are called *fission* reactions. From figure 7.2 you can see that the strontium and xenon nuclei that emerge from the ^{235}U reaction each have a higher binding energy than the parent uranium nucleus. This is why energy is released by the ^{235}U fission reaction. All fission reactions start on the right side of the binding energy curve, and the reaction product nuclei are higher up the curve (i.e., closer to the most stable nucleus, that of ^{56}Fe). A fission reaction to the left of ^{56}Fe cannot occur because splitting an atom of, say, ^{16}O will result in smaller nuclei (to the left of ^{16}O on the curve) that have lower binding energies. Making this happen would consume energy instead of releasing it.

However, there is a way to extract energy from nuclei that lie to the left of ^{56}Fe, and that is by combining two or more small nuclei to make one larger nucleus. Reactions in which this occurs are called *fusion* reactions. From the shape of the binding curve you can see that a great deal of energy will be released if, say, hydrogen nuclei can somehow be forced together to make a helium nucleus. Fusion reactions are much more difficult to generate and sustain than are fission reactions, however, as we will see. Consequently, all commercial nuclear power plants today make use of fission.

RADIOACTIVITY

Of the four fundamental forces of nature, we now come to the fourth one: the weak nuclear reaction. This shows itself in the *radioactive* property of many naturally occurring elements. Radioactivity happens in the nucleus of an atom when that nucleus spits out a particle, usually an X-ray, a gamma ray, an electron, or a helium nucleus. At the same time, energy is also released, causing the radioactive material to glow—as with radium—or become hot. The weak force is not nearly as powerful as the strong force (perhaps you might have reached that conclusion unaided), and yet the particles released can be very energetic.

All the fissionable material used in nuclear reactors are naturally radioactive, as are most of the fission products. For example, we have seen that ^{235}U splits into ^{95}Sr and ^{139}Xe. Both these decay products are radioactive, and so they spontaneously decay, releasing an electron and some energy. On average, the strontium will decay within about 24 seconds, and the xenon within about 40 seconds. The radioactive decay products may themselves be radioactive, in which case the sequence of spontaneous decays continues until a stable (nonradioactive) nucleus is formed.

The problem, as we all know, is that the particles spat out by radioactivity can be a health hazard. Radioactive by-products are an inevitable consequence of fission reactions used to generate nuclear power. Some radioisotopes are extremely dangerous, and it is the health risk posed by these radioactive products, not the risk of nuclear power plants turning into nuclear bombs, that worries opponents of nuclear power.

Reservations about Reserves

With the basics of nuclear physics laid out, we now turn to an examination of nuclear power. Let us begin by examining the fuel supplies available for nuclear power plants. As with all aspects of the nuclear debate, the data concerning reserves are contested, with wildly different estimates from different sources. I will not attempt to be comprehensive in my survey of the literature. Instead, I will point out a few articles that are representative and air some aspects of the subject that are important to know before we can make an intelligent appraisal.

The first fact that comes to light, almost as soon as we begin to investigate the fuel that is needed for fission reactors, is that identifying the fuel itself is slippery. This in no short measure accounts for a significant part of the discrepancies concerning reserves. True, the most commonly used nuclear fuel is ^{235}U, which forms 0.7% of natural uranium (most of the rest is ^{238}U). True, uranium is a fairly common metal in the earth's crust (at 2.8 ppm it is about as common as tin or zinc). So far, so good. The problem with defining nuclear fuel is that other isotopes of uranium or other elements such as plutonium and thorium can be used as fuel, and (slipperiest of all) that certain nuclear reactors can transmute one isotope or element into another, thus creating more nuclear fuel than they use up. Let me explain.

The bare facts concerning ^{235}U resources are these. The current world nuclear reactor capacity is 375 GW (over half of it in three countries—the United States, France and Japan). Most of this comes from uranium fuel, of which 68,000 tons are used annually. The largest reserves of recoverable uranium are in Australia (1.67 million tons), Kazakhstan (651,000 tons), Canada (485,000 tons), and Russia (480,000 tons); the world total is 5.4 million tons. Do the math, and you will arrive at the figure of 80 years' supply.

From 25,000–100,000 tons of ore, 200 tons of uranium oxide (called *yellow cake*) is extracted by grinding the ore to a powder and leaching it with sulfuric acid. Most reactors require enriched uranium: they require a fuel that contains 3%–5% ^{235}U rather than the natural 0.7%. Enrichment is achieved by converting the oxide into a gas and passing it through a series of centrifuges, thus separating the lighter ^{235}U from the heavier and much more abundant ^{238}U. Through this process, 200 tons of yellow cake is reduced to about 25 tons of enriched uranium, which is pressed into pellets and placed into fuel rods suitable for inserting into reactor chambers. These 25 tons will fuel a 1-GW reactor for a year.[4]

The 80-year supply figure becomes less certain—indeed, it almost dissolves in the air—when we dig a little deeper and consider three more facts. First, as with all mined resources, the amount of recoverable uranium depends upon its price (the figures just given assume a uranium price of $130/kg). If the price doubles, then the estimated recoverable uranium reserves increase by a factor of 10. There was little exploration for uranium between 1985 and 2003, for reasons I discuss later, and so few new reserves were identified. Since 2003 several have been found, with the result that the global supply increased by 15% between 2005 and 2007. Also, of course, the number of years' supply that remains assumes that we continue to use up uranium at the current rate. This is unlikely to be true. For example, between 1980 and 2003 world nuclear capacity increased by a factor of 3.6, while fuel consumption increased by a factor of 2.5 (the difference is due to increased plant efficiency).

Second, the purity of uranium ore varies widely from region to region, and this affects the cost of recovery. Thus, low-grade ore contains less than 0.1% uranium, whereas the fraction for high-grade ore is 2%. (In Canada there are some very high-grade deposits, consisting of 20% uranium.)

Third, and most important, the current method of burning fuel is very inefficient. The *thermal reactors* in use today extract only 2% of the energy content of the uranium fuel. The so-called spent fuel that emerges from these reactors thus contains 98% of the energy. There is another type of reactor—the *fast breeder*—that can recycle the spent fuel so that all of it can be used. If these reactors become widely adopted, then our fuel reserves increase by a factor of 50. We will look at the different types of fission reactors in the next section.

The strange nature of nuclear fuel is evident after further consideration. Currently, 78% of the uranium fuel that is used in reactors comes from mining. This is the primary supply, but there are secondary sources of uranium from commercial stockpiles and decommissioned nuclear weapons. (Weapons-grade uranium is highly enriched—pure—and so goes a long way when diluted as power plant fuel. Swords into plowshares, almost.) Add to these the currently small amount of recycled uranium and plutonium, obtained by reprocessing spent fuel, and we arrive at the 22% figure for secondary supply.

There are unconventional sources of uranium, analogous to the tar sands for oil reserves. Some 4 *billion* tons of uranium is contained in the oceans of the world. For the foreseeable future this source will remain uneconomical to recover. Nuclear reactors can also use other fuels; we will see that the Canadian Deuterium Uranium (CANDU) reactor is capable of burning thorium as well as uranium. India is investing in thorium reactors. Thorium is three to four times more abundant in the earth's crust than is uranium.

Quite apart from these genuine complications, there are less creditable reasons for the differing estimates of nuclear fuel resources. For example, proponents and opponents of nuclear power can disagree on facts that are readily checked. Thus, we find several sources reporting the discovery of new uranium deposits at the same time that another source claims that no new deposits have been found since 1980. Clearly, there is some wishful thinking going on, or burying of heads in sand. The evidence suggests to me that proved recoverable uranium reserves will last considerably longer than any fossil fuel reserves.[5]

Fission Reactor Technology

Most nuclear power plants today are of second- or third-generation design. When nuclear power first came into the world in the 1950s, there were many different design types, some safer than others, some more efficient than others, some more economical than others. Nuclear engineers proceeded by trial and error, and by the second generation they had pared down the designs to a few types. Of these, pressurized water reactors are the most common because their design is relatively simple and is considered (by nuclear engineers) to be very safe. They are compact (having evolved in the United States from nuclear submarine power plants, developed in the Cold War).They cannot turn into nuclear bombs. For the very same reason, they also cannot be turned into *fast breeder* reactors. I will unpack fast breeders soon enough—they may play a very important part of the world's nuclear power future—but first we will look at PWRs and two other common reactor designs, the CANDU reactor and the Russian RBMK reactor.

In order to understand the plethora of nuclear reactor designs, and to have some insight into their advantages and disadvantages, you need an appreciation of one or two technical details. First, let's tackle the important distinction between fast and slow neutrons.

A slow neutron (usually referred to as a *thermal neutron*) is about a thousand times more likely than a fast neutron to split a nucleus of ^{235}U. This fact of nature greatly affects nuclear reactor design. Recall that bombarding ^{235}U with neutrons produces more neutrons—but these are fast neutrons. To ensure that the chain reaction happens, it is necessary to slow the neutrons down. In nuclear reactors, this slowing down of neutrons is achieved with a *moderator*. A common, inexpensive, and very effective moderator is ordinary water. For reasons that will soon become apparent, the water that falls from the sky or comes out of your kitchen faucet is known to nuclear physicists as *light water*. The most common type of nuclear power plant in the Western world—the pressurized water reactor—uses light water as a moderator.

PRESSURIZED WATER REACTORS

In figure 7.3 we see how pressurized water reactors (PWRs) work.[6] Figure 7.3a shows control rods suspended above the nuclear reaction chamber.

(a)

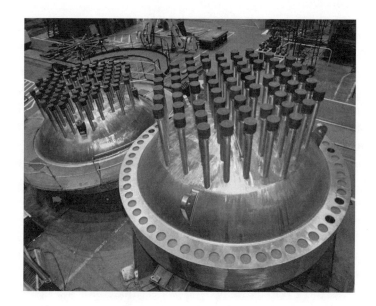

(b)

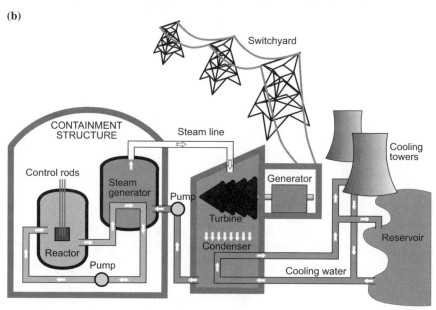

FIG. 7.3. Pressurized water reactor. (a) The top of a reactor chamber, showing control rods. (b) Schematic of PWR operation. The primary loop, which enters the reaction chamber, contains the light water moderator under high pressure (so that it does not boil). The secondary loop contains water that is converted to steam, which powers the generator. An additional loop condenses this steam. (These last two loops are common to many types of generator.) *(a) NRC image; (b) figure from the U.S. Energy Information Administration.*

Control rods contain material that captures (absorbs) neutrons so as to dampen down or halt entirely the chain reaction process, depending on how far the control rods are inserted into the reaction chamber. A basic safety feature of many Western reactor designs is this: the control rods are held up by electromagnets; if electrical power is lost for whatever reason, the electromagnets lose power, and the control rods fall under gravity into the reaction chamber, shutting down the reaction. This idea of safe shutdown as a response to a failed reactor component—in this case, electrical power—is known as *failsafe*.[7]

Figure 7.3b shows a PWR design schematic. Note that there are three loops of circulating fluid that transfer heat. First, water inside the reactor is heated up before being pumped to a steam generator. This water is under high pressure and is enclosed within the pipe; it is separated from the water used to generate steam. It is the water that circulates through the reactor which gives the PWR its name: this water moderates the neutrons that emerge from ^{235}U decay so that a chain reaction occurs, resulting in the generation of heat. Unfortunately, water is not just a moderator of neutrons; it is also an absorber of neutrons. This means that the water tends to sop up neutrons and dampen the chain reaction—the opposite of what we want. To counter this absorption effect, PWRs use enriched uranium oxide fuel—that is, fuel that has an enhanced content of fissionable ^{235}U (3%–5% instead of the naturally occurring 0.7%). Enriched uranium produces more neutrons than unenriched uranium, and these extra neutrons counter the absorption by water, so that a sustainable chain reaction occurs.

Heat from the pressurized water of the first loop is transferred to water in the second loop, converting it into steam. This steam drives turbines, which generate electricity. The Rankine cycle (discussed in chapter 2) is most efficient when the steam that comes out of the turbine condenses back into water, which is then returned to the steam generator. To ensure that the steam condenses, a third loop of water is employed. Cooling water is circulated between cooling towers and the turbine condenser. Since the water in the second and third loops is isolated from the nuclear reaction chamber, it is not bombarded with radiation and does not become radioactive.

The fuel in PWRs needs to be replaced every 18 to 24 months. It has significantly depleted its stock of ^{235}U and has built up fission products that are useless as fuel (with a notable exception soon to be discussed) and are often highly radioactive. Typically, a third of the fuel is replaced each time. What happens to the spent fuel is important and has generated much controversy, as we will see later in the chapter. The significant operational aspect of PWR refueling is that the reactor has to be shut down during the refueling process. During this complex procedure, the power plant is not generating electricity, so PWR refueling adds to running costs.

CANDU REACTORS

The ubiquitous PWR is found in many of the 30 countries that employ nuclear power (though mostly in the United States, France, and Japan). One disadvantage of the PWR is its need for enriched fuel. The Canadian Deuterium Uranium (CANDU) reactor does not have this problem because it uses natural (0.7% ^{235}U content) uranium dioxide fuel. In fact, it is not a picky eater at all: it will devour almost anything fissionable, including the spent fuel from PWRs. It has proved to be a popular design in half a dozen countries outside of Canada. (At home, it is the sole type of nuclear reactor used to generate electricity.) How does it achieve this fuel trick, and given this advantage, why aren't all nuclear nations using CANDU reactors?

CANDU reactors use *heavy water* as a primary loop coolant instead of light water.[8] In heavy water, two deuterium (D) atoms replace the hydrogen atoms, so the chemical formula is D_2O instead of H_2O. Deuterium is chemically the same as hydrogen, but its nuclear properties differ because it has a neutron as well as a proton in the nucleus. The physical properties of heavy water are very slightly different from those of light water because the two extra neutrons in a molecule make it 11% heavier.[9] The nuclear properties differ significantly; in particular, heavy water does not absorb neutrons nearly so readily as light water (it is already neutron-rich), and yet it still slows them down. Consequently, it is a very good coolant.

The low neutron absorption of its moderator means that a CANDU reactor works with uranium oxide fuel that is not enriched. It also works with MOX (mixed oxide—plutonium as well as uranium) fuel. Spent fuel is MOX because the abundant ^{238}U absorbs neutrons to become ^{239}Pu, which, after being hit by a second neutron, fissions to release energy. (In a light water reactor, plutonium is a waste product because there are not enough neutrons to turn it into a fuel.) Plutonium is the fuel of choice for nuclear weapons, and many nuclear power plant designs produce plutonium as a waste product in spent fuel, thus acting as a potential factory for nuclear bomb fissionable material. The CANDU reactor consumes its plutonium and so is considered safe to export to countries that may have one eye on developing nuclear weapons as well as nuclear power.

The fact that the CANDU reactor does not require enriched fuel also helps make it safe for export because enrichment technology is another key component of weapons development. If a nation can enrich its nuclear power plant fuel from 0.7% to the 3% or 5% needed for light water reactors, then it can keep on going and enrich the fuel to 90% for weapons-grade uranium. Much of the fuss over Iran's stated desire, at the time of writing, to develop an indigenous nuclear power capability concerns its enrichment facilities; can we trust the Iranian government not to keep the centrifuges spinning until they produce weapons-grade material? With CANDU reactors, there is no need for enrichment; thus, countries which accept CANDU cannot easily use their nuclear technology to develop weapons.[10] In fact, one of the main reasons the Canadians chose the heavy water route was because, in the early years after World War II, they did not have the capability to enrich nuclear fuel.

Apart from the saving in enrichment costs, the use of heavy water saves on waste disposal. Light water reactors contain radioactive plutonium as well as fission products in their waste fuel, which is expensive to store safely; heavy water reactors consume much more of their fuel, so less spent fuel is produced, mitigating the waste management costs—and the produced waste is less toxic. On the other hand, the heavy water coolant is much more expensive than light water. The safety features of the CANDU design also add expense; it does not have the failsafe mechanism of light water reactors, but additional redundant heat-dump mechanisms make it a safe design.[11]

RBMK REACTORS

An old Soviet design of PWR, the RBMK reactor is cheap, powerful—and unstable. It uses light water as a coolant but, uniquely, employs graphite as a moderator. The fuel for this reactor is slightly enriched, to about 2%, and the result (pre-Chernobyl) was a reactor with a runaway capability: an increase in reactor temperature would cause the coolant water to boil, thus reducing its density and its ability to absorb neutrons, thus increasing neutron flux and core temperature, and so on. In the aftermath of Chernobyl, this design was put under a microscope, as you may imagine, and numerous other faults were found. There was no secure containment of the reactor core worthy of the name (a reinforced concrete shell acted as a radiation shield); the response of the control rods to dangerously high temperatures was sluggish; emergency cooling mechanisms were flawed. Most seriously, operator expertise was lacking, in part because design flaws were not discussed—the RBMK design was a state secret—and criticism of the design was not encouraged. In fact, operator error led to the disaster at Chernobyl. An increase in core temperature during a training exercise led to a steam explosion that exposed hot graphite to the open air. The graphite burned and spewed radioactivity high into the atmosphere.

Post-Chernobyl, many RBMK reactors were shut down, or were replaced by other types after an RBMK reactor was decommissioned. Even so, there are still 11 of these reactors, each producing 925 MW, operating in Russia. Their safety features have been significantly beefed up since Chernobyl, with faster scram time (shutdown, via control rod insertion); installation of a fast-acting emergency protection system; improvements to control rods and core cooling; and so on. An image of bolting horses and barn doors comes to mind.

FAST BREEDER REACTORS

Fast breeder reactors (FBRs) are currently in the prototype stage but are sufficiently well developed for most of the nations with an interest in nuclear power to have one or two such research facilities. We have seen that reactors are often characterized by the type of coolant that they use because this determines their key operating characteristics, and in particular, it influences the type of fuel they use and the spent fuel that they produce. Fast breeders are, in this respect, like CANDUs on steroids: they have a high neutron flux because they operate with a coolant (often a liquid metal such as sodium) that does not absorb many neutrons. Consequently, they do not need enriched uranium fuel—indeed, their main fuel is the abundant, non-fissile ^{238}U. The main purpose of fast breeders is to create fuel for other nuclear reactors, not to generate power. A nucleus of ^{238}U absorbs a fast neutron and becomes ^{239}Pu. Plutonium can be used as a fuel (as well as to make nuclear weapons).

Herein lies a technical problem. The FBR creates ^{239}Pu from ^{238}U, and so ^{239}Pu builds up in the reactor core. The fuel rods cannot be left inside the reactor until all the ^{238}U has turned into ^{239}Pu because the ^{239}Pu will be struck by fast neutrons and will fission (this happens by design in CANDU reactors). Such fissioning will deplete the ^{239}Pu. There comes a point—a density of ^{239}Pu among the ^{238}U—when for every plutonium nucleus that is created by the neutron flux, another one is destroyed. Before this point is reached, the fuel is removed from the reactor core and is sent to a *reprocessing* plant, where the ^{239}Pu is separated and sent elsewhere to be burned as nuclear fuel.

The remaining fuel is returned to the FBR, and the process begins anew. Some FBRs can create as much as 30% more fuel than they use. In this way, the very common but nonfissile ^{238}U is converted into usable fuel for nuclear reactors.

A number of countries have built experimental or prototype fast breeder reactors for the purpose of generating fuel: these include the United States, the United Kingdom, France, Russia, India, Japan, and Germany. The implementation of FBRs is technically different from that of commercial nuclear reactors, and at present they are not economically competitive. (They become so when the cost of uranium is high.) Spent fuel reprocessing was halted in the United States under the Carter administration. The Germans lost their enthusiasm for the idea and have shut down their facility. After Fukushima, they lost their enthusiasm for nuclear power of any sort and are seeking other solutions to the energy problem, as we will see in the next chapter. As a consequence of Fukushima, the Japanese also have begun to waiver in their support for nuclear power. At the time of writing, 49 of the 54 Japanese nuclear power plants appear to be idle, and it seems that the $12 billion Japanese FBR program is in jeopardy. Japan is likely to switch from a policy of recycling nuclear fuel to a *once-through* policy (common to many countries, including the United States) of burying their spent fuel. Recall that spent fuel still contains some 98% of its original recoverable energy. The earthquake and tsunami of March 2011 is casting a long shadow on nuclear policy.[12]

Fusion Technology

There is far less to say about fusion technology than about fission technology for the simple reason that there are no commercial fusion reactors in existence and probably will not be before the end of this century.

Recall from the binding energy curve of figure 7.2 that nuclear energy can be extracted by combining light nuclei such as hydrogen as well as by splitting heavy nuclei such as uranium. The attraction of fusion is that it releases even more energy than does fission. It does not involve dangerous radioactive fuel (and hydrogen can be extracted from water inexpensively) or byproducts, and we know that fusion reactors are stable (the sun has been around for a long time). However, it has proved much more difficult to find a way to combine nuclei than to split them. A star forces hydrogen nuclei together by virtue of the enormous gravitational pressure generated by its constituent matter. On earth we have to find a different technique.

Research facilities in Europe and the United States concentrate on two possible methods for confining, as a prelude to combining, nuclei. The nuclei to be confined are in the form of plasma—a hot, ionized phase of matter that cannot simply be put in a strong container such as a conventional fission reactor vessel. The problems are enormous. Current ideas for confinement involve laser beams or magnetic bottles. So far, the best result obtained for a fusion reaction was fleeting: a 10-MW reaction confined for half a second. Even this result, however, used up more power than it produced. Fusion power researchers believe they can reach the breakeven point (where power out matches power in) in the next decade and hope that this advance will spur development of what could be a clean and potentially inexpensive source of electrical power (the original "too cheap to meter" claim for nuclear-generated electrical power from the early, optimistic days after World War II).[13]

Fusion power may well be clean and, eventually, inexpensive. But not in this century. We as a species need another source of power that will carry us through to the day when fusion power solves the problem for millennia to come.[14]

The Nuclear World

There are currently 30 members of the nuclear club—the group of nations that employ nuclear fission to generate electricity for their citizens—and several more (such as Iran) that are knocking at the door. The club membership is shown in figure 7.4, where the enthusiasm of each member for this controversial technology is displayed graphically.

We have seen that most, but by no means all, of the nuclear reactors used by these nations are of the PWR type. There are other proven designs,[15] and new reactors are being built with new (so-called "generation IV") technology. The new generation will produce less toxic waste and cheaper electricity and will, it is claimed, be safer. Generation I nuclear power was born in a period of optimism, but then the Cold War, with its threat of global nuclear war, and a series of more or less serious nuclear accidents (prominently at Three Mile Island in Pennsylvania in 1979 and Chernobyl in Ukraine in 1986), turned public support into fear and opposition. The generation II

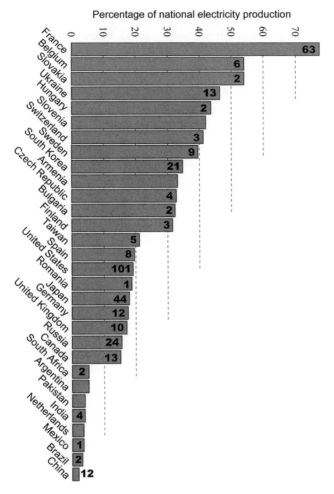

Percentage of national electricity production

France 63
Belgium 6
Slovakia 2
Ukraine 13
Hungary 2
Slovenia
Switzerland 3
Sweden 9
South Korea 21
Armenia
Czech Republic 4
Bulgaria 2
Finland 3
Taiwan 5
Spain 8
United States 101
Romania 1
Japan 44
Germany 12
United Kingdom 10
Russia 24
Canada 13
South Africa 2
Argentina
Pakistan
India 4
Netherlands
Mexico 1
Brazil 2
China 12

FIG. 7.4. Nuclear power around the world. The percentage of each nuclear nation's electrical capacity that is provided by nuclear plants ranges from less than 10% to more than
70%. The bold numbers show each country's nuclear capacity in gigawatts (if there is no
bold number, the total capacity is under 1 GW). Only France, Belgium, and Slovakia generate more than half their electrical power from nuclear sources; the United States, France,
and Japan account for more than half of the total nuclear capacity. *Data from EIA (2010a).*

reactors, with the glaring exception of RBMK, work well enough; and generation III reactors work well, judging by results (electricity prices, safety,
production costs). Attitudes began to change. There always was and always
will be a hard core of opponents and proponents who are unlikely to budge,
but the floating public reaction became less hardened against the nuclear

energy option by the turn of the millennium. There remained nagging doubts, however, about nuclear waste disposal.

Nuclear reprocessing to more efficiently extract energy from nuclear fuel and reduce the volume and toxicity of nuclear waste is more expensive than the once-through policy adopted by most nuclear nations today. (Only the United Kingdom, France, Russia, and India currently reprocess spent fuel from civilian nuclear power plants.) Reprocessing and the associated fast breeder reactors are in many ways the future of nuclear power, but people worry about its potential for nuclear proliferation, as we have seen.[16] In the United States, President Bush reversed Carter's decision and supported a reprocessing program; President Obama has backed away from Bush's decision.

Few people are happy with the once-through approach because of the large volume and long half-life of the radioactive wastes that it produces. The nascent or perhaps still-born Yucca Mountain nuclear waste repository in Nevada is an interesting symbol of our ambivalence toward the subject. Opponents of nuclear power do not want the repository to be used because of the large volumes of high-level radioactive waste that would be placed there. Proponents of nuclear reprocessing point out that the Yucca Mountain facility would not need to be so big if spent nuclear fuel were reprocessed; there would be much less waste to bury than with the current once-through policy. Funding for Yucca Mountain was withdrawn in 2009, and so the United States now has no permanent site for its growing nuclear waste. Another example of this ambivalence (dithering, perhaps): Canada also uses once-through reactor technology and therefore buries a lot of nuclear waste. However, the Canadians' waste is recoverable, should they decide in the future to reprocess it.

After the end of the Cold War, when uranium prices dropped and the need for reprocessing became less, the once-through approach became less and less expensive compared with reprocessing. Yet with the inevitable end of cheap fossil fuels looming, with the possibility of minimizing nuclear waste through FBRs and reprocessing, and with increasing public concern over global warming, greenhouse gases, and environmental degradation in general, it seemed in the early noughties (as the decade 2000–2009 has come to be called) as if there might soon be a nuclear renaissance—a reappraisal of the role to be played by nuclear power in our future.[17] Nuclear power can get us off fossil fuels, it produces much less carbon dioxide (none, directly), and properly handled, it is environmentally friendly.

Then came Fukushima. The next three paragraphs summarize the immediate short-term effects of this terrible disaster; in the medium term, it has turned public opinion against nuclear power once again. (Consequently, the International Energy Agency has halved its prediction of new nuclear capacity that will be installed across the world by 2035.)[18]

In the afternoon of March 11, 2011, a magnitude 9.0 earthquake—one of the most powerful ever recorded—occurred some 40 miles off the coast of Tohoku, northeastern Honshu, at a depth of 12 miles. It was felt over a wide area and later triggered over a thousand aftershocks. More immediately, it led to a very large tsunami that hit the east coast of Japan with waves up to 130 feet high. Some of these traveled as much as six miles inland. It has subsequently been estimated that, within the course of an hour, nearly 20,000 people were killed (15,844 are known to have been killed, and 3,450 are missing). In addition, there were 26,992 injuries. The disruption was considerable: 130,000 buildings collapsed, and another 690,000 were damaged; 4.4 million households lost electricity, and 1.5 million lost water. The total cost of the disaster has been estimated by the World Bank as $235 billion—the most expensive natural disaster in history.[19]

The six nuclear reactors at the Fukushima-Daiichi coastal plant were hit by the tsunami. Two of these reactors had been shut down prior to the earthquake for scheduled maintenance; the other four shut themselves down automatically in response to the earthquake. The tsunami severed the connection of the plant to the national electricity grid and flooded the generators that had kicked in to provide emergency backup power. Once the generators stopped working, there was no power to circulate coolant, so the four online reactors overheated. Reactors 1, 2, and 3 underwent meltdown; hydrogen gas explosions spread radioactive material over a wide area. Eventually, seawater was used to cool the reactors; it also rendered them permanently inoperable.

In response to the radiation (which would amount to 10% of the radiation released at Chernobyl),[20] people were evacuated from areas within 20 km (12 miles) of the Fukushima plant. The radiation release means that nearly 300,000 people in Japan will, over the course of their lifetimes, have received a total external dose of 41 mSv. (I discuss dosage units in the next section; this dosage is about 100 times the dose we normally receive each year just from the food we eat.) And 2,200 people will receive doses that are 20 to 100 times higher than this. A preliminary study by Rensselaer Polytechnic Institute estimated that cancer mortality rates in the region have been boosted 0.001% above the natural mortality rate, so that approximately 100 people will die from the effects of radiation.[21] The Japanese government banned food products (milk, lettuce, spinach) from the affected regions. It has been criticized for its handling of the disaster and has admitted shortcomings in its response.[22]

The Japanese will be living with the consequences of the earthquake and tsunami for a long time. The rest of the world will be deciding how to respond to the nuclear accident at Fukushima, and living with the consequences of that decision, for at least as long.

Statistically Safe?

Radiation effects are statistical and merit a separate section. It is reasonable to place such a discussion here, though the perception of risk influences the energy debate quite generally—not just the nuclear debate. I will have to rein myself in, be warned, because this subject is something of a burr under my saddle.

DOSAGE DIFFICULTIES

Be suspicious of claims about radiation levels and dosage, for several reasons. There are lots of units for measuring both, and comparisons are difficult. Thus, radiation comes in three basic types—alpha, beta, and gamma—and these have very different effects on human health. Some radiation is delivered from a compact source, while some is diffuse. Some radioactive substances, such as potassium, concentrate within the human body, whereas others are spread more or less evenly. Two exposures to the same amount and type of radiation may have very different health consequences, if one is a high dose of short duration and the other is a low dose of longer exposure (the first—the high dose—is more dangerous).

A few statements can be made that are clear-cut. The radiation levels and dosages received at Hiroshima and Nagasaki, close to the bomb sites, were higher and more dangerous than those inside the reactor site at Chernobyl. The latter were catastrophic for the cleanup crew who went into the Chernobyl reactor site immediately after the explosion. Doses further afield are more problematic to interpret. Radiation levels downwind of Chernobyl varied with distance, of course. For what it is worth, they were 190 microsieverts (μSv) in parts of Germany but only 0.2 μSv in more distant Portugal. We each receive about 400 μSv per year from the food we eat. The Portuguese received an average dose of radiation from Chernobyl that is "equivalent" to the dose they would have gotten from eating two bananas. (I put "equivalent" in quotes because the nature of the radiation was different.)

The problems of interpreting health damage from radiation are largely a result of its statistical nature. We humans find it difficult to properly understand statistical events and very difficult indeed to include them in law.[23]

PERCEPTION OF RISK

It is difficult to have a sensible debate about statistical matters because the notion of statistical confidence is so slippery. There is no such thing as statistical safety, for example. There is a very small probability that you will be struck by lightning during your lifetime, but it might happen; and if it does, it might seriously damage your health. How should we respond to the risk? There is a similar probability of winning a major prize in a lottery. Do we react to these two possibilities as if the likelihood of these events is the same? No.

The prompt and terrible consequence of the Japanese tsunami of March 2011 was the death of over 19,000 people. Perhaps 100 will die over the next several decades as a consequence of radiation exposure from Fukushima-Daiichi. In response, the Japanese are seriously considering their nuclear power future. Are they seriously considering removing their coastal population to higher ground? No. The number of deaths resulting from the 1979 partial meltdown at Three Mile Island is probably in the single figures (with zero being the most likely); 30 years later, in 2009, the number of deaths in the same state (Pennsylvania) from automobile accidents was 1,256. Three Mile Island has significantly influenced national energy policy for a third of a century. Many people called for a ban on nuclear power; how many are calling for a ban on automobiles?

Of course, we should do everything we can to minimize the adverse health consequences of any useful technology. Lessons should be learned from disasters—whether they be in coal mines, airplanes, or nuclear power plants—to make technology safer. It seems to me that a wide section of the population regard nuclear energy as a different kind of risk from other technologies, a risk that is uniquely loaded with the potential for planet-wide disaster. Certainly, if 30,797 people had died in a single nuclear power plant apocalypse in 2009, it is likely that all nuclear power plants would have been shut down and nuclear power would be permanently taken off the table as a contender for our future energy needs. Had that number of people died as a result of football stadiums collapsing during 1979, we might expect that new stadium designs would be improved and existing stadium safety would have been beefed up across the country—but that people would still go to football games. In fact, the number 30,797 refers to the number of automobile fatalities in the United States in 1979.[24]

Presumably, our irrational response to nuclear risk is due to history (Hiroshima and Nagasaki influence our attitude about Fukushima) and to the etymological connection with nuclear explosions. Yet the connection be-

---

CONCENTRATED POWER

Recall the power area density calculations of table 4.1. These show how much power is generated per unit area of plant site. In fact, the detailed calculation involves more than the plant site: it takes into account offsite areas that are necessary for the plant to operate (e.g., coal mines for a coal-fired power station). Power area density is an important consideration when we consider a technology for large-scale power generation—say, as a replacement for oil when global supplies run out. We can perform the same calculation for nuclear power that we applied earlier for other power generation technologies. The results, you might anticipate, will be very favorable, given that the energy density of uranium is four orders of magnitude greater than that of coal, but it is not so, as we will see. Nevertheless, the nuclear power industry has a large power area density which matches the best of other industries.

Consider the Torness nuclear power plant in lowland Scotland, built on a 200-acre site, with two reactors each supplying 625 MW to the national electricity grid. The nominal power area density of this site is thus 1,500 W/m². This calculation does not take into account the area needed for waste disposal and reprocessing or the land that is taken up by the uranium mines which supply Torness with its basic fuel. Or consider the giant Bruce nuclear power plant in Ontario, Canada. It generates 6,232 MW on a 2,300-acre site that includes a heavy water plant (required for CANDU reactor coolant, you may recall) and waste storage facilities. This plant thus has a more representative power area density figure of 650 W/m². A similar calculation for three American nuclear power plants chosen at random (Beaver Valley, Pennsylvania; Brunswick, North Carolina; and Byron, Illinois—I chose plants whose names begin with a *B*) yields an average power area density of 400 W/m².

These calculations show that nuclear power has a power area density that is far higher than that of any renewable resource and is comparable to coal and natural gas figures.

---

tween nuclear power and nuclear explosives is almost as tenuous as the connection between plastic containers and plastic explosives. The nuclear power industry likes to point out the silliness of our attitudes by demonstrating that, as we saw earlier, more radiation is released into the environment by burning coal than by operating a nuclear power plant of the same capacity.[25]

We can do our best to make technology safe and can estimate our success statistically. That is, we can analyze past events mathematically. Such analyses show the likely number of deaths that have been or will be caused by nuclear power plants, coal power plants, oil power plants, and the other power generation technologies. Similar analyses can be made of environmental damage caused by various energy technologies. So long as we use the same criteria for all cases, the comparisons based upon such analyses are valid. Similarly, we cannot single out nuclear power for the potential damage it might cause if some unlikely disaster were to occur in the future. Arguments that we can envisage mass death from a terrible mistake due to nuclear power or pharmaceutical research, and that given the risk we should consequently avoid both, are equally invalid. Some people argue that we cannot anticipate all possibilities and therefore should not proceed with nuclear power because it is intrinsically impossible to make it perfectly safe. True, but the same can be said of any other power technology, or of medical research for that matter.

Nuclear Screening

As with other power industries, it is difficult to pin down the costs of nuclear power development and operation absolutely, so the best approach is, I think, to consider the relative costs of those aspects that can be reasonably estimated.[26]

A quick example will serve to illustrate the relative economic merits of nuclear power. A *screening curve* brings to light one important aspect of power production technology, and it has the advantage that it can be compared across technologies. As was true of load duration curves in chapter 1, the data can be presented graphically so that the relevant information can be understood at a glance. In this case, the graph is of cost per unit of generated power plotted against the load factor (capacity factor—the fraction of peak production rate at which a power plant operates). Thus, if a power plant outputs power at its peak rate for three weeks and then shuts down for a week, it has a load factor of 75% for the four-week period.

In figure 7.5 we see a typical curve (this one happens to be for the United Kingdom). Such a curve is useful for determining cost effectiveness across industries at a given time and place. The slopes and positions of the lines will change over decades as technology influences the different industries in different ways, and the lines will differ by country. This one shows, for example, that when plants are producing power at a very low rate (below 20% of full capacity) open-cycle gas turbine technology is the most economical. As the load factor increases (from 20% to about 70% capacity), combined-cycle gas turbines are the best choice. OCGT and CCGT are two types of natural gas-fired power plants. When power needs to be produced near peak capacity (above about 70% in this example), nuclear power is the most cost-effective. We can understand why the screening curve for nuclear power is nearly flat: there is a high capital cost, but the running costs are quite low, independent of load factor and with low fuel costs. Screening curves help utility companies and national governments decide what mix of generating technologies to invest in and operate.[27]

The reserves of nuclear fuels are greater than those of fossil fuels and may last for thousands of years if fast breeder reactors and spent fuel reprocessing are

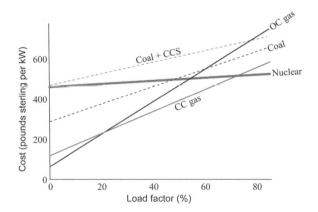

FIG. 7.5. Screening curves for production costs of nuclear, coal (with and without carbon capture and storage [CCS]), and gas power (open-cycle [OC] and combined-cycle [CC] turbines) in the United Kingdom. *Data from Green, Hu, and Vasilakos (2011).*

adopted. This approach will also mitigate the problem of storing radioactive waste. There are several designs of fission reactors; with the exception of the Soviet RBMK design, these are as safe as or safer than other power generation technologies when judged by likely number of fatalities. The adoption of fusion technology for power generation will eliminate concerns over radiation, but this technology is many decades in the future. Attitudes toward nuclear power are different from attitudes toward other sources of power, partly because the first application of nuclear technology took the form of devastating bombs and partly because the effects of radioactivity are statistical, and statistical risk is poorly understood by many people.

8

HERE COMES THE SUN

The main renewable energy sources are wind power and solar power—the subjects with which we begin this final chapter. We start off with a survey of wind power technology and capabilities. You might reasonably wonder why wind power should fall under this chapter title. In fact, winds are driven by solar power.[1] We then consider again the possibilities of solar power, which we first glimpsed in the introduction. In the second part of the chapter we look at the mix of power generating technologies which, it seems to me, offer the best prospects for the future.

The Answer, My Friend, Is Blowin' in the Wind

Wind energy is all around us and is free, in the sense that nature delivers it right to our doorstep, whether we want it or not. Harnessing that energy is not free—quite the reverse—and yet power from the wind is being increasingly realized. The sails of wind power research and development are filling, so to speak. Wind is a clean, renewable energy source; exploiting it does not generate greenhouse gases or cause chemical pollution.

From the windmills of old, we have over the last several decades developed wind turbines of various shapes and sizes. The vertical-axis wind turbine (VAWT) has been around since the 1920s and has found several applications, mostly small scale. The most widespread and familiar of these is the cup anemometer, used to measure wind speed. There are also larger eggbeater-like designs that generate power. These have one advantage over the much more common horizontal-axis wind turbine (HAWT) design: they do not need to be turned into the wind. This is a distinct advantage in areas where the wind direction is highly variable and unreliable—for example, on the tops of urban buildings. Generally though, VAWT designs are uncommon because they rotate slowly and therefore generate only a little power.[2]

HAWT designs usually have three blades; they vary in size from 200-W micro-turbines to large megawatt turbines. The micro-turbines, with blades 0.5–1 m long, are used for generating power for sailboats in harbor, RVs, electric fences, and water pumps (one of the original windmill applications), and are used by hobbyists. Stepping up in size we find mini-turbines that generate a few kilowatts; these have blades up to 3.5 m long and are used to power remote cabins or households.

I will concentrate here on commercial turbines, designed to generate power for industries or, more usually, for an electricity grid. The blades for these turbines range from large—20 ft— to gigantic, as long as 165 ft (fig. 8.1). The giants can generate serious amounts of electrical power. They stand atop 300-ft towers and are usually grouped to form a wind-power plant, or *wind farm*.

As with many technologies, scale affects economics and efficiency. Micro- and mini-turbines are 20%–25% efficient, whereas large turbines can be up to 40% efficient. The power of a turbine depends on the swept area of its blades, so it increases as the square of blade length. Power also depends sensitively on the winds, increasing as the cube of wind speed. (You may

FIG. 8.1. Transportation of a large wind turbine blade in Texas. *Photo by Alexi Kostibas.*

recall from chapter 2 that we showed this dependence on blade length and wind speed to be the case for windmills; the calculation is in the appendix.) This is why scaling up a wind turbine yields a disproportionate increase in power. Another reason is the *wind profile*—the increase in wind speed with height above the ground. Friction with the ground slows the wind; at increasing heights there is a consequent increase in its speed. Atmospheric scientists assume this speed to increase as the one-seventh power of height.[3] So, doubling the height of a turbine tower from, say, 60 feet to 120 feet increases mean wind speed by 10%. Assuming that turbine blade length increases proportional to tower height, we find that the wind power impinging upon the turbine blades increases by about 540%. All of this means that turbines for wind farms have to be large to be economical.

Installation costs for wind turbines are very high. For a small turbine, reckon upon spending $230 per square foot of swept area, and perhaps $93 per square foot for a large turbine. Recall that available wind power and turbine efficiency both increase with turbine size, and you can see why size matters when it comes to wind turbine economics. Maintenance costs are high because the wear and tear on turbines is considerable. (A small turbine can be expected to last about 10 years.) A large turbine is turned into the wind to maximize the swept area and also to ensure that the blades do not have to deal with turbulence caused by its tower (called *mast wake*). The turning of large turbines is via computer-controlled motors (small windmills use the wind, as we saw in fig. 2.3c). High winds, and especially gusting winds, are the enemy of wind turbines; pitch control (*feathering*) to avoid dangerously high speeds is incorporated in modern designs, and blade-tip sensors are included to activate brakes.

For onshore wind turbines, siting is very important. We have seen that it is better for a turbine to be high up, but not all high sites are equally good. A tower is superior to a building roof because of turbulence caused by the building. Ridgelines are often adopted, partly because they are high and partly because a fluid-dynamics effect accelerates the wind that flows over such topographical features. The turbines must be spaced so they don't steal each other's air; the rule of thumb is a 12–20 blade length separation between turbines, and it is up to 30 blade lengths for large wind farm turbines. Wind speed and profile vary with surface roughness and topography as well as upwind obstacles. All of these factors favor an offshore location for wind turbines, where the wind in some locations is also more predictable and reliable, resulting in a greater load factor for offshore wind farms. The

United States is a leader in onshore wind power technology and implementation, whereas Europe leads in offshore wind power.

The wind is iconically temperamental and quixotic, and this unreliability is the main headache for wind farm designers and operators.[4] For example, meteorological records are not particularly helpful in determining the best site for a wind farm because meteorological anemometers are usually placed less than 30 feet above the surface. Big turbine towers can be 10 times higher than this, where the wind conditions can be quite different. Designers must therefore gather their own detailed information about the wind characteristics at a potential site as part of the planning process. Given the nature of wind, these data are statistical, and statistics forms a significant aspect of the integration process in which wind farm power is added to an electricity grid, as we will see.

The largest wind-power plant in the world is currently the Roscoe Wind Farm in Roscoe, Texas. This facility consists of 627 wind turbines spread over an area of 150 square miles and has an installed capacity of 780 MW. The construction cost was one billion dollars. We can see from these figures that the capital cost per kilowatt is high and that the power area density is low (see the sidebar "Roscoe"). However, power area density considerations need not be a significant factor because the land between turbines on a wind farm site can still be used for other purposes (such as agriculture, as we see in fig. 8.2). The same cannot be said for a nuclear power site or a fossil fuel plant.

In the United States wind power now accounts for about 3% of electricity generating capacity (44 GW). This figure—the second highest, behind China —is likely to be out of date by the time you read it because wind power generation is rapidly increasing in the United States. Thus, in 2010 the size of American wind-power facilities increased 15%; a quarter of all new power that was added to the electricity grids during that year came from wind farms. The Department of Energy anticipates that wind power's share of electricity generation will increase to around 20% by 2030 (and that electricity demand will increase by 39%). In the studies that make this prediction, assumptions are made about technological advances that will reduce cost and improve performance, but these assumptions seem to be cautious and reasonable. Currently 85,000 people work in the U.S. "windustry" (let me coin this word—it seems a natural contraction of "wind industry"). This number, and the political and economic clout of wind power employees, is likely to increase over the next few decades.

Europe boasts the world's largest offshore wind farm, a 307-MW facility

FIG. 8.2. A wind farm in lower Saxony, Germany. Most of the land beneath the turbine towers is used for agriculture. *Photo by Philip May.*

near Walney, off the northwest coast of England. Europe's current offshore capacity is 2.4 GW (the Danes and the Germans are very keen on this particular technology, for reasons that are easily appreciated by anyone who has visited the coasts of northern Europe), but there are proposals to increase this capacity within the foreseeable future to 100 GW.

These numbers are impressive. They suggest that if wind power can be made cost-effective (we have seen that capital costs are high—and are especially so for offshore wind farms), it might play a significant role in providing us with power from, say, 2030. It is likely, however, that most of our electrical power will not be provided by the wind because wind is erratic. Both water and gas enter turbines at a constant rate, by design, so that the turbine rotors turn at a constant rate and power is supplied to an electricity grid at a constant level. For wind turbines, the input power is very variable, so there is a considerable engineering overhead in minimizing the effects of variable output power from the turbines.

Consider the Mid-Atlantic Bight, a stretch of water off the New England coast that has been studied as a possible site for an offshore wind farm. The wind patterns there have been investigated in detail. At a given location the wind speed might be 10 mph at 10 a.m. and 25 mph at 2 p.m. In this

ROSCOE

The 780 MW of the Roscoe Wind Farm, a facility that is spread over an area of 400 km², yields a power area density of 2 W/m², which is pitiful compared to that of nuclear power or natural gas. However, because the wind farm land can be used for other purposes, power area density is not a significant consideration. A billion dollar cost for 780 MW amounts to a capital investment of $1.28 per watt. Assuming 35% efficiency, we can infer an average wind speed of about 10 m/s for a 40-m blade length. The minimum speed at which wind turbines work is about 3 m/s (7 mph). Given the power consumption of the average Texan, the Roscoe Wind Farm supplies about 570,000 people with electrical power—say 250,000 homes. The same wind farm operating in the same wind in Germany would supply 850,000 homes.

instance, the power represented by the afternoon wind is nearly 16 times that of the morning wind (recalling that power increases as the cube of wind speed). Wind turbine operators like to keep the blades rotating at speeds in the range 10–22 rpm. This restricted range—the "comfort zone" for turbine operation—eases the task of the sophisticated gearboxes and power converters that interface with the electric grid. Nevertheless, not all the fluctuation in power levels caused by wind speed variation can be compensated for by controlling the blades to maintain a comfortable rotation rate.

For micro-turbines it is possible to even out the variability in supplied power by temporarily storing power in batteries and releasing it when the wind speed drops, but for large turbines this is not practical. In fact, it is not necessary or even desirable to do so. Windustry engineers know that statistics can come to their aid. A large wind farm with many turbines spread over a large area, such as Roscoe, will provide a total output power that fluctuates less, as a fraction of the average, than the output power of any single turbine. This is because, over a large area, a gust of wind in one spot may boost the power output of a few nearby turbines while a lull in another spot will reduce the power of turbines in that area. The net effect is a decrease in the fluctuation in output power for the wind farm as a whole. (More precisely: the fractional fluctuation of the wind farm's output will be less than the fractional fluctuation of any turbine within it.)

Even with this statistical calming due to large numbers of turbines, the power level that a wind farm provides to an electrical grid is more variable than that provided by other technologies. What is worse: it is not predictable. We saw in chapter 1 that grid operators face considerable challenges in coping with a variable and quixotic electricity *demand*; with wind power as part of their power source they face similar uncertainties on the *supply* side. Some windustry designers have suggested having wind power work in concert with hydro, which we know can respond quickly to rapid changes in demand. The idea here is that the hydro plants would adjust their output in response to changes in wind power supply, thus evening out the power level. Others argue that this is not necessary; statistical studies show that today the net effect of large and widely distributed wind farms imposes a small and tolerable regulation burden of some 50 cents per megawatt-hour of electricity generated by wind power. The reason is the smoothing effect of large numbers of turbines, here writ larger over an entire grid and not just over a single wind farm. In the dry phraseology of one Department of Energy report, "Aggregation reduces variability."[5]

It is important to note that this 50-cent burden applies because the current wind power contribution is only 3% of the national electricity supply. As the wind power share increases, regulation costs will rise. Integrating wind power into a grid will require a great deal of new infrastructure. Wind farms are often remote (certainly, those offshore are well off the beaten track of existing power transmission lines). The calming effect of large numbers will reduce wind power variability but will not eliminate it. The Department of Energy anticipates that further regulation of power will impose a high cost—perhaps as much as $5 per megawatt-hour if wind power constitutes 20% of grid capacity. (And with further regulation a utility operator will have to include the effects on and of other power sources working with wind power.) The reason is the inexact nature of weather forecasting. We are not able to predict, one day ahead, what the wind conditions will be. Consequently, peak-load generators will have to be started up every day (either to add power to the grid or to store it by, for instance, pumping water uphill) even if they are not used. This is expensive, and unless we learn how to forecast local wind conditions accurately and precisely across large areas (which seems very unlikely), the uncertainty in wind power supply will probably restrict the amount of wind power that we can control and utilize.

Stand a Little Less between Me and the Sun

I mentioned Solar Energy Generating Systems (SEGS), the world's largest solar energy plant, in a note to the introduction. This facility generates 354 MW at nine sites spread across the Mojave Desert in California. There are also large solar energy plants in China (where the Golmud Solar Park, generating 200 MW, is the largest single site in the world), Germany, Spain, Italy, Canada, and the Ukraine. The power generated by all solar plants across the world, however, is less than 1% of the total, though it is rising: some 40 GW was added in 2010.

There are two types of solar plants. One concentrates the beams of sunlight that are intercepted by large arrays of mirrors, focusing them on a central tank where water is boiled; this type of facility is shown in figure 8.3. The boiled water creates steam, which drives a turbine in the conventional manner. Such *concentrated solar power* or *thermal solar power* plants use the energy of sunlight indirectly—via boiled water and steam turbines—to generate electricity. Another variant of this type does not use mirrors; instead, it circulates water through solar panels. Thermal solar power plants account for only 1,100 MW of generated electricity. This is less than 1% of the power generated by the other type of solar plant—the *photovoltaic* (PV) solar power plant (fig. 8.4).

PV plants convert the energy of sunlight directly into DC electrical energy by making use of the photovoltaic effect.[6] The solar panels inside which this conversion is made have until recently been very costly to manufacture, making solar power uneconomical. Over the last two decades, however, the cost of solar panels has dropped to the point where solar electricity costs (about $1.50 per watt in 2010) now approach those of conventional sources.

The development of solar power, perhaps more than any other power technology, has been directly influenced by economies of scale. Nowhere is this more apparent than in Germany, the Western capital of solar power. As a result of the Fukushima disaster, the German government decided to phase out nuclear power and concentrate on renewable sources. For some time it had already been encouraging renewables—wind, tidal, and solar power—and the decision to abandon nuclear power left it with no choice but to pursue renewable energy full tilt. The government did what many Western governments do to influence the direction of technological development, only more so: it introduced a carbon tax to reduce carbon dioxide emissions (making fossil fuels more expensive) and *feed-in tariffs*—FITs—

FIG. 8.3. Focusing mirror at a solar site in the French Pyrenees. *Photo by Björn Appel.*

FIG. 8.4. An array of photovoltaic (PV) solar panels at a 1-MW power plant in Death Valley, California. *Photo by the author.*

to encourage expansion of solar panel production (making solar panels less expensive).

FITs are subsidies that oblige utilities to buy electricity from solar power plants, as well as other renewable power sources, even though solar power is more expensive than electricity from other sources. These tariffs also provide incentives to homeowners to put solar panels on their roofs and to owners of larger buildings and facilities (e.g., apartment buildings, swimming pools, etc.) to invest in solar heating; see figure 8.5 for a Spanish example. The idea is to create a market for solar panels, thus increasing the demand for them, thus increasing production levels, and finally (through economies of scale) driving down the price of each panel. As the price falls, the FITs are reduced. They are removed altogether when prices for solar electricity achieve parity with the price of electricity that is generated from conventional sources.

Many other countries apply FITs—Italy in particular, following the German model; the large Canadian solar plant at Sarnia, Ontario, has also benefited from this type of subsidy. In Germany, however, such subsidies have been applied on a grand scale and have been, it seems, too successful.

FIG. 8.5. An urban PV array attached to the side of a building in Spain. *Photo by Chixoy.*

Though Germany is not the obvious country in which to pursue solar power (cloud-shrouded Germany receives only a third the sunlight of Spain, for example), Germans now derive 3.2% of their electricity from sunlight—a figure that is several times the world average. In 2011 some 7.5 GW of solar power was added to the German electricity grid, and Germany now has 25 GW of solar power—40% of the world's total. No other country in Europe, and only China and the United States in the wider world, approach this solar capacity. Yet the very success of the FIT program has caused the German government to slash the subsidy by 30%: it is costing too much as solar panel companies proliferate. Consequently, several high-profile solar companies have gone to the wall. One ambitious project in the United States that has been mothballed because of the turmoil in German solar power circles is the Blythe Solar Power Project, a 500-MW-capacity plant in California. This plant was under construction when the construction company went bankrupt as a result of the insolvency of its German parent.

Solar power costs are all about scale of production. The trouble is that PV panels are difficult to make. At present there are no thin-film PV cells that are simultaneously nontoxic, durable, and inexpensive. Yet the government subsidies have resulted in PV-generated electricity prices that are tantalizingly close to parity with other sources.[7] The effects of scale are emphasized by comparison with the plummeting costs of digital memory chips and microprocessors, by Jenny Chase, of the market research firm Bloomberg New Energy Finance, in London: "The photovoltaics industry follows an experience curve similar to that described by Moore's Law. Every time the production of crystalline silicon [solar power] cells doubles, the cost per unit comes down by 20%–25%. That has been a very clear path since the 1970s."[8]

Another example of solar power innovation that is largely a result of economics: Saudi Arabia is investing heavily in solar for domestic use and hopes that by 2020 solar power will supply 10% of the nation's electricity. This makes sense, given that the country consists of a lot of desert (though desert dust can be a real problem for PV panels). Can we assume that the Saudis are looking beyond oil, to a future in which they are still a significant energy provider? Perhaps, but this is not the immediate motivation. It turns out that the high price of oil (which will very likely remain high as oil becomes scarce) makes it worthwhile: they are selling abroad the oil that would have been consumed domestically.

Like wind power, solar power is intermittent, though not so unpredictable. (We can be pretty certain that it will be unavailable at night, for example.) It has a low power area density (less than 10 W/m²; see table 4.1). There are some environmental issues: PVs contain heavy metals such as lead, mercury, and cadmium. More significantly in the short term, there are serious doubters in Germany. The big energy companies there can be expected to oppose solar power, and are doing so cogently, by pointing out how Germany would do better to pursue offshore wind power and hydropower and leave solar power to countries such as Spain, which will eventually generate it for less cost per kilowatt than is possible in Germany. Another fly in the ointment is the attitude of neighboring Poland. The Poles see an opportunity to sell cheap power to the Germans, given the (perhaps transitory) high cost of German energy resulting from the policy of subsidizing renewable sources. The irony here is that the Polish energy will be nuclear—and so nuclear power may grow in Poland as a result of the German decision, post-Fukushima, to abandon its own nuclear power program.

A Distant Sun

A long way in our future we might imagine sunlight providing much of the world's power. If we placed satellite reflectors in space, the energy of sunlight could be gathered continuously so that the problems of intermittency would not apply. It is a useful exercise to investigate this possibility, if only to show that we are some distance (in technology and time, as well as space) from satellite power generators.

A lot of large light-gathering satellites will be required to provide electricity for a power grid. (Figure 8.6 shows a NASA illustration of one possible type of satellite power generator.)[9] Each will be in geostationary orbit 22,000 miles above a point on the earth's surface. It will receive mostly visible light (because that is what the sun emits) and convert this power into microwaves, which are transmitted down to the ground. The transmitting array is always directly above the rectenna on the surface (because the array is in geostationary orbit). Why convert to microwaves? Because visible light is absorbed quite strongly by the atmosphere and is absorbed or reflected by clouds and rain, whereas for certain microwave frequencies these objects represent no barrier at all, as any radar engineer will tell you.

The microwave radiation reaching the ground will have to be limited in density because it will irradiate a large surface area and may otherwise be a health hazard. In the appendix, I provide a Fermi-like back-of-the-envelope calculation that shows how much power each of these arrays could provide,

FIG. 8.6. NASA illustration of a satellite microwave power generator. This is a small low-earth-orbit Suntower generator, which absorbs sunlight and redirects the power as microwaves to a receiver on earth. Because it is in low earth orbit, it cannot beam power down to the same receiver all the time; its orbital speed is too fast.

given sensible parameters and including a safe limit on surface radiation density. This calculation suggests that an array in geosynchronous orbit should be about 0.8 miles (1.3 km) in diameter, whereas the ground rectenna beneath it should have a 3.4-mile diameter.[10] The electric power delivered to a power grid on earth, from each of these array-rectenna pairs (let me call the pair a "satellite generator"), would be about 735 MW. Today, the power consumed by humanity is 14–20 TW, as we saw in the introduction; this total could be provided by 20,000 satellite generators. Such power would be clean, everlasting, and (because nobody owns sunlight) free.

Free? Whoa—let's back up the horses a little. Check out the sidebar entitled "Skyrocketing Costs," and you will see that this whole idea of satellite power generators is, for now, pie-in-the-sky. The purpose of this little exercise is not to nay-say or ridicule the concept as economically and energetically infeasible but, instead, to show that it is unrealistic today and for the next few decades, at least—until some key technological advances are made. The satellite generators need to be made more robust so that they can stand up better to the hostile "weather" found in space; the launch cost per kilo-

SKYROCKETING COSTS

Let us see what happens if we decide to solve the world's energy problems today in a single bound, with satellite generators beaming solar power down to earth. The hostile environment of space (lots of cosmic rays and very fast dust particles) takes a great toll on satellites: a satellite is doing well if it survives for a decade. At such an attrition rate, the 20,000 satellites we will need in geostationary orbit will have to be replaced at the rate of 2,000 each year. A minimum weight estimate for each satellite comes in at 2,500 tons.* A commercial heavy-lift rocket such as Falcon Heavy delivers a payload into geostationary transfer orbit that constitutes 1.35% of its weight. The math tells us, therefore, that replacing dead satellites will require launching 390 million tons of stuff into space each year. The energy required by heavy-lift rockets to overcome gravity amounts to 130,000,000,000,000,000,000 J per year, which is equivalent to an average power consumption of 4 TW. So, about a quarter of the total world power output would be needed just to maintain the required number of satellites in orbit.

What about economic cost? A Falcon Heavy can carry 19 tons of payload, so that about a quarter million of these rockets would need to be launched each year (that's one every two minutes). The cost per kilogram of payload is $2,200 (much lower than with other launch vehicles). The annual cost of replacing burned-out satellites is then about $10 trillion—about equal to half the entire U.S. GDP.† And then there is the environmental cost of all those launches . . .

---

*I obtain this by assuming that the 1,300-m-diameter array is on average 1 mm thick and made from a material that has twice the density of water.
†The 10-year lifetime for space equipment is from Brekke (2004).

---

gram of payload needs to be brought down— in terms of both dollars and fuel. The hard facts of the matter tell us that, until these advances get here, large-scale power generation from space is not going to happen.

There are other satellite-rectenna configurations that may be more sensible than the one I have considered here, but the bottom line will have the same number of digits.[11] Laser light, instead of microwaves, has been considered as the carrier of power from space to earth, though here, the problem of atmospheric absorption and reflection becomes severe. However we cut and dice the problem, it will be decades or centuries before large-scale space-based solar power becomes feasible.

So Where Do We Go from Here?

In two hundred years' time, I believe, we will have electricity that is too cheap to meter. It will probably be provided by nuclear fusion, not solar satellites. In that distant future we will have overcome the considerable technical problems that must be solved to render fusion feasible and economical. By the dawn of the twenty-second century we may have expensive but functional fusion reactors. Assuming that humans are as flawed then as they are now—the safest assumption I will make in this book—we can expect fusion disasters along the way. During our journey toward too-cheap-to-meter electricity, fusion reactors will blow up or leak, killing or injuring a few unfortunate workers and spilling debris far and wide—harmless hydrogen and helium, mostly.

In 5 years—and probably in 50 years—we will still be burning dirty old coal (rebranded as "clean coal," which it is, but only when compared with dirty coal) and oil and gas. These fossil fuels pollute heavily and will run out way before fusion power becomes feasible. (We may still be burning coal in the year 2100 because there are more coal reserves than reserves of gas and oil, or we may still be burning gas because it is less damaging to health and to the environment.) So the question is: how do we generate the power we will need for the century or more that forms the transition period from fossils to fusion? Another question that exercises many minds: can renewable energy replace other sources entirely? These are the central questions of the long-running energy debate, and they are of prime importance for the future of our species.

Natural gas is the least polluting of the fossil fuels and has significant conventional reserves. By all means, let us use it until the conventional supply runs out. What about shale gas? We will probably be stuck with it, just as we will be stuck with shale oil, tar sands oil, and coal. Even if we see environmental disaster looming, we will probably not be able to stop using these fossil fuels until the last cubic foot is burned, for several reasons. The many decades of technological advances in the field of fossil fuel extraction, processing, and use give the fossil fuel industries considerable inertia; it will take time to wean cars and power plants off them. It will also take time to wean governments off fossil fuels because the coal, oil, and gas industries are part of the fabric of nations—they have been for over a century—and the special interests of these powerful industries will find many advocates in positions of political influence. Oil and gas, and coal, will become more and

GENERATING COSTS AND EFFICIENCIES

In the figure I present—perhaps in a novel way—data from the Energy Information Administration. The EIA has calculated the total system-levelized cost of energy from different sectors—in other words, the cost per megawatt-hour of energy produced, taking into account capital expenditure for construction, as well as operating and maintenance costs and performance and fuel costs. (Not included are loan costs and other purely financial factors, which depend much more on the characteristics of individual projects than upon technology.) The costs are in 2009 U.S. dollars, assuming that a plant enters service in 2016.

I have chosen the most economical sources of renewable power (thermal and not PV solar power, onshore and not offshore wind power). At the same time I have chosen the most ecologically friendly fossil fuel plants (coal and gas power plants that employ CCS—carbon capture and storage). In this sense I am buffing up renewables and dragging fossil fuels through the mud, making the best economic case for the former and the worst for the latter. The levelized cost for each technology is plotted against the capacity factor—the fraction of full operating capacity which the technology can generate, on average. Thus, the best place for a power generation technology to be on this graph is the top left corner.

The numbers inside the circles represent power plant efficiency as a percentage. The heat engines (gas, coal, nuclear) all operate at 40%–45% efficiency, meaning that this fraction of the fuel energy content is converted into useful energy

more expensive as supplies dwindle, and this will eventually cause them to disappear as fuels, but the process will be glacial.

Currently, 80% of our energy is obtained from fossil fuels. Given that they will fade away sometime in the last half of the present century (my guess), what source of energy will we turn to? Hydropower is a good start. We have seen that it is clean, renewable, competitive, and safe. We have also seen that it is limited by geography and geology and is therefore a minor component of the energy mix for most nations of the world. Clearly, for fortunate countries such as Norway, the continued use of hydro is a no-brainer—why would they turn to any other source of power? But for the world as a whole, hydropower will always be a small slice of the pie. I will guess that hydro will contribute 5% to the world's power at the time fossil fuels run out; this guess is double the current contribution and won't be far wrong.[12]

Because of the scale of investment in wind power and the level currently reached by wind power technology, it is reasonable to guess that the 20%

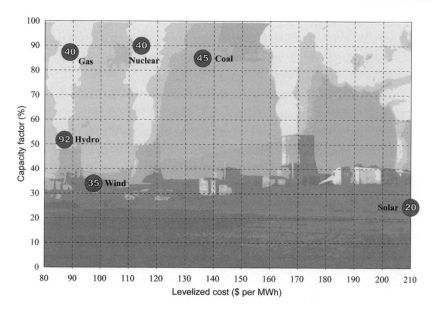

Capacity factor vs. levelized cost for six power-generation technologies. The number inside each circle represents typical technology efficiency (as a percentage). *Data from EIA (2010d).*

(e.g., electrical energy). Hydro shines brighter than the heat engines, while solar power is left in the dark.

target for wind power in the United States by the year 2030 is attainable. Wind is a universal phenomenon, unlike flowing water; it is thus possible that this figure may be attained by the world as a whole by mid-century and later. We have seen that if wind power is boosted to a higher share of the power that supplies a grid, integration problems arise that are expensive to fix. Perhaps technology will find a way around that, but perhaps not—let's stick with 20% for wind power.

There has been a great deal of hype for solar power over the past few decades, and it looks like it is paying off at least in part. Investment in solar power technology is reducing the high cost of solar panels to the point that they are beginning to become competitive economically, even without subsidies. (They are not quite there yet; see the sidebar "Generating Costs and Efficiencies.") We can anticipate that this might occur sometime about mid-century. So how big a slice of the pie will solar power take? It is an intermittent source, like wind power, but unlike wind power it is also very geographically uneven. It makes sense for Spain or Arizona to invest in solar power, but not for Britain or (dare I say) Germany. We saw in the introduction that the power area density of solar power is small. This may change—and costs may indeed drop—but I will guess a sober 10% for solar power by mid-century.

Hydro, wind power, and solar power between them, according to my guesstimates, will provide 35% of our energy by the second half of the present century. Other renewables such as geothermal or biofuel are going to contribute negligibly, on the global scale, for different reasons.[13] Thus, it seems to me that renewable technologies are going to provide us with barely a third of the power we need when fossil fuels have run their course; they are not going to be able to fully replace fossil fuels. This fraction may increase with the decades, but we will need a lot more power before this increase happens. Or will we? Could our governments not convince us to consume less power—perhaps enough to close the gap? After all, Americans and Canadians consume about three times as much power per capita as do the citizens of Western Europe, for no very great difference in living standards. If we can significantly reduce the fuel consumption of our cars, as we have over the last four decades, why can we not do the same for other aspects of our lives?[14]

It ain't gonna happen. Maybe a bad conscience (or, more probable, a bad economy) will lead to a reduction in the electrical power consumed by each North American—but for every one of these people there are seven people from China and India who are itching to consume electricity at the rate that First Worlders currently do. Because of this, and because the population of earth will increase, it is likely that both per capita and total energy and power requirements across the world will grow and grow.

Without fossil fuels and with renewables unable to bridge the gap, we will have no choice in the medium term but to rely upon nuclear fission. We have seen that it is reasonably economical, reasonably clean, and reasonably safe. There are no guarantees, of course, except that human error will generate more accidents in all sectors of the power generation industry (people have died on wind farms). There will be more nuclear accidents, especially if we are reckless enough (oh, how keen is hindsight) to place nuclear reactors along the shoreline in a tsunami zone or near geological fault lines.

With care, the number of deaths or injuries from nuclear power will continue to be small when compared with the number of deaths (by radiation, even) from coal-fired power plants—or tanning salons. Tanning salons?

Well, people are beginning to wake up to the fact that tanning salons deliver electromagnetic radiation to pale customers who pay to receive it and that such radiation may be damaging to health. Again, this is a matter of perceived risk versus reward. It seems that for some folk a good-looking tan is worth an exposure to ultraviolet radiation. Similarly, people who worry about the radiation from smart meters may complain about it to their friends via cell phone—holding a source of microwave radiation to their ears.

I'm sorry if this hobbyhorse of mine is getting ridden into the ground, but the irrationality of our response to unseen dangers is going to cost us dearly if we persist in denying ourselves the use of nuclear power because of it. That a technologically sophisticated nation such as Germany should reject nuclear power because of Fukushima, while continuing to emit vast amounts of dangerous pollutants from coal-fired power plants, would be funny if it wasn't so sad. Even if we don't like nuclear power, common sense and historical data tell us that it is relatively safe (in the Western world) and that we will kill more people in the long run by not using it than by using it.

One of the advantages of nuclear power is that it is not tied to particular regions of the globe where reserves lie. The sources of uranium are many and widespread, and it is unlikely that we would ever go to war over nuclear power plants, as we have done over oilfields. (Nuclear reprocessing and uranium enrichment by aspiring nations is a different matter, but our decision to pursue or not pursue nuclear power domestically is not a part of that dogfight.) There may be wars over energy resources in the coming decades until fusion power (or maybe even solar satellite power) is feasible and economical, and there will be casualties from the energy sources that we do use (in particular, respiratory diseases from fossil fuels), but nuclear power will contribute only a small part to both these causes of human unhappiness. The bottom line is this: even if we push the more lovable renewable alternatives as far as they will go, we have no choice but to employ nuclear fission power.

Wind farms play an increasing role in our energy budget. Technological advances and sophisticated integration strategies will be required before wind power can gain the 20% share that is foreseen by 2030. Thermal solar power (focusing sunlight to boil water) is becoming competitive, but PV solar power (direct conversion of sunlight via the photovoltaic effect) is still significantly more expensive than other sources of electricity, though feed-in tariffs are rapidly bringing the cost down. Unless satellite solar power becomes feasible, it is unlikely that we will be able to derive a significant amount of power from the sun.

In the distant future we will have fusion power as a source of cheap and clean power. Before fusion becomes available, we will run out of fossil fuels and will need to utilize fission as a transitional source of power. (Renewable sources of power will provide only a third of our requirements.) We must learn to overcome the natural but irrational fear of nuclear power, while working to make it as safe as we can.

AFTERWORD

When two people disagree fundamentally, we know that at least one of them must be wrong. When experts disagree, the rest of us can only make educated guesses as to what is right. In this book I have quoted several experts on the subject of energy and future energy sources and have investigated the subject from the semidetached point of view of a scientifically trained outsider (outside the energy industry but an interested bystander who will have to live with the consequences of the choices made—as we all will). Here is Professor Marten Hoffert, an expert (if 30 years' experience in the field of alternative energy technology and global environmental change makes him an expert): "Terrorism doesn't threaten the viability of the heart of our high-technology lifestyle, but energy really does." Here is Alan MacFadzean, an oil expert we met in chapter 5: "The world as a whole is not going to stop using oil and gas any time soon and will pay whatever is needed to get the stuff because we all know that if we don't, our entire civilisation will fall apart. And that is not hyperbole."[1]

We can make the same point less dramatically, with the quiet certitude of scientific (some would say dry) data to back us up, by considering again figure 2.9. This plot shows a strong positive correlation between the GDP of a nation and its per capita power consumption. The natural assumption is that people in wealthy countries have the disposable income to spend on power, to heat their homes, to drive big cars on long vacations, and so on. However, correlation says nothing about cause and effect, so the causal relationship might just as well be the other way round. Suppose that high power consumption generates national wealth (for example, because enterprising people drive their goods farther to market).[2] If this is indeed the case, then we cannot seriously contemplate a significant reduction in power consumption in the face of reducing fossil fuels and increasing environmental pollution. To do so would result in a significant reduction in the quality of our lives and those of our children. The recent financial troubles of our banking system (an avoidable farce, if ever there was one) shows that our children have little enough to thank us for: we have borrowed off them for more than a generation to pay for extravagances, not for sound infrastructure or a secure future.

We have the chance—indeed, the obligation—to think about, and develop, our future sources of energy and power. It behooves us to think carefully and choose wisely.

APPENDIX

The following sections provide additional, more technical information on some of the topics in the text. The emphasis here is on quick, easy-to-see estimation, so I am happy to settle for approximate answers. Simplifications are made to ease calculation. Of course, the art in doing this sort of thing is to choose simplifications that don't throw away the essential physics.

Compressed Air Storage

Let's assume that when air is compressed, the compression is isothermal—there is no change to the temperature. In this case Boyle's law applies:

$$PV = P_0 V_0. \tag{A.1}$$

That is, the product of pressure and volume of the air doesn't change. In (A.1) the subscript 0 indicates initial value. The energy required to compress air to a volume V_1 is

$$W = \int_{V_0}^{V_1} P dV. \tag{A.2}$$

Substituting from (A.1) gives us

$$W = P_0 V_0 \ln\left(\frac{P_0}{P_1}\right). \tag{A.3}$$

In (A.3) P_1 is the pressure after compression of the air to volume V_1. Note that W is negative because we have to put energy in. The energy that can be extracted from the compressed air is $-W$. We will assume that a compressed air storage facility is optimized to store the maximum possible energy. Elementary calculus tells us, from (A.3), that this occurs when $P_1 = eP_0$, i.e., when the final pressure is about 2.7 times the initial pressure, in which case the total energy stored in the system is

$$W_{max} \approx P_1 V_1. \tag{A.4}$$

The wobbly equal sign, $\approx$, indicates that the equation is expected to be only approximately true, because of my simplifying approximations. Equation (A.4) is something we can work with; it gives us the stored energy in terms of parameters we know—the final pressure and the fixed volume of the storage facility.

To estimate the time that a power station can operate by using the compressed air, we can multiply W_{max} by the plant efficiency and divide by the output power. This is how I arrived at the numbers in chapter 1.

Overshot Water Wheel Power

The power of an overshot water wheel derives from the rate at which water flows over the wheel and the drop in height of the water as it flows over the wheel. The energy of a given mass, m, of water is mostly gravitational potential: $E = mgh$, where g is the acceleration due to gravity and h is the height through which the water drops; this height is twice the wheel radius R (i.e., $h = 2R$). The mass flow rate is $\dot{m} = \rho A v$; here, the dot indicates time derivative, ρ is water density, A is headrace cross-sectional area, and v is water speed. Thus, a simple estimate of the overshot wheel power is

$$P = \dot{E} = \dot{m}gh = \rho A g h v \epsilon, \tag{A.5}$$

where ϵ is water wheel efficiency. I made use of equation (A.5) to arrive at the water wheel power noted in chapter 2.

Windmill Power

A rough estimate of windmill power can be found in a manner similar to that I have just applied to water wheels. The main difference is that wind energy is kinetic and is given by $E = \frac{1}{2}mv^2$, where v is wind speed. The mass flow rate is $\dot{m} = \rho A v$, as with water wheels, but now ρ is air density and $A = \pi R^2$ is the cross-sectional area of the windmill sails, where R is sail length. Thus, windmill power is estimated to be $P = (\pi/2)\, \rho R^2 v^3 \epsilon$, where ϵ is windmill efficiency. This is the formula I drew on in chapter 2.

Let's put in some numbers, to see the kind of power we can expect. Air density is about 1.24 kg/m³; a large windmill might have sails 90 feet long (27 m). Say wind speed is a moderate 6 m/s; efficiency is low, at 5%. Such a windmill will generate about 15 kW.

Energy Efficiency for the Kingston Power Plant

The Tennessee Valley Authority states that its Kingston coal-fired power plant, located west of Knoxville, generates 10 billion kWh of electrical energy per year and uses 14,000 tons of coal per day. Putting these numbers together gives us a fuel consumption rate of 2,000 kWh per ton. Typical coal has an energy density of 6,670 kWh/ton, so the efficiency of Kingston is about 30%. This figure falls within the normal range for coal-fired plants but is below the anticipated efficiency (45%) for the next generation of plants.[1]

Pipeline Engineering: Calculations for TAPS

The Trans-Alaska Pipeline System can transport as much as 2 million barrels of crude per day, corresponding to a volume flow rate of $Q = 3.68$ m³/s. The diameter of the pipe is 48 inches ($d = 1.22$ m); we can determine the average speed of the oil through the pipeline as $v = 3.0$ m/s (about 7 mph). Thus, traversing the 800-mile distance ($L = 1,287$ km) from Prudhoe Bay to Port Valdez takes 5 days. The crude must be forced down the pipes, because friction acts against the flow, and so we ask: how much power is needed to maintain flow in TAPS?

The simplest equation that describes fluid flow of this sort is the Darcy equation, which can be written in the form shown below. (In applying the Darcy formula here, we are making a number of simplifications and approximations. For example, we are ignoring the effects of change of height: TAPS snakes its way over mountains.)

$$P = \frac{8\,\rho f L Q^3}{\pi^2 d^5} \, . \tag{A.6}$$

In (A.6) ρ is the density of crude oil (call it 860 kg/m³), and f is a dimensionless friction factor that is variable, depending upon pipe surface roughness and diameter, but here averages about 0.013. Note that the required power, P, is proportional to pipeline length, increases more rapidly with oil flow rate, and is most sensitive of all to pipe diameter. Plugging in numbers, we find that to force crude oil along the pipeline system against friction requires about 215 MW (or 288,000 hp). In fact, published data show that TAPS is provided with 12 pumping stations, each of which consists of three 13,500-hp pumps, leading to a total pumping power of 292,000 hp (218 MW). The very close agreement with my simple estimation is probably fortuitous: what should be ballpark agreement is significantly better than that, but I'm not complaining.

It is natural to ask how much oil needs to be burned to provide this power. Given an average energy density of 37 MJ per liter for Alaskan crude, knowing the required power and transit time along the pipeline, and assuming a pump efficiency of 60%, we can readily show that about 0.3% of the crude oil transported via TAPS is used to overcome friction. Stated differently: of the 1287 km of oil in the pipeline, some 3.6 km is needed to power the flow.

Now let us see how much pressure must be exerted by the pumps to facilitate this flow. To this end we turn to another version of the Darcy equation:

$$p = \frac{\rho f L v^2}{2d},$$

(A.7)

where p is pressure. Again plugging in numbers for TAPS, we obtain a pressure of 53 MPa, or 530 atmospheres. This is a very high pressure to be maintained in a half-inch-diameter pipe and is why the pipeline is provided with 36 pumps and not one huge pump. You can see from equation (A.7) that pressure is proportional to length; with the pipeline divided into smaller lengths the pressure is distributed and reduced to a more manageable 15 atm.

Switching to a quite different aspect of TAPS, we can show why the pipes must be equipped with large lateral sliders (fig. 5.2). Thermal expansion can cause a length of pipe to "buckle," and the pipeline base must be constructed to allow for this lateral movement. The following diagram shows the geometry and notation. The thermal expansion of steel is characterized by a linear expansion coefficient of $c = 0.000013$ C^{-1}, meaning that for each 1°C rise in temperature, a rod of steel will increase in length by 13 parts per million. This sounds small, but it leads to significant buckling. The buckling distance is

$$x \approx \ell \sqrt{2c \, \Delta T},$$

(A.8)

where 2ℓ is the distance between two fixed point of the pipeline, shown in the figure, and ΔT is the temperature change. Thus, if the pipe is straight when the ambient temperature is -35°C (-31°F, which is not unusual for an Alaskan winter), then in summer when the temperature is 15°C (59°F), a 200-m section of pipe will buckle 3.6 m, say 12 feet, in the middle if the end points are fixed. You can see why sliders are needed to allow for movement due to thermal expansion.

How Far on a Tank of Gas?

The aerodynamic drag force acting on a car is $f_{drag} = \frac{1}{2} c_D A \rho v^2$, where c_D is drag coefficient for the car (which depends on car shape), A is the cross-sectional area of the car, ρ is the density of air, and v is car speed. For a typical small car traveling at 55 mph (25 m/s) this force is about 200 N. At this speed, aerodynamic drag accounts for about 60% of the total force acting on the car (the rest is engine and

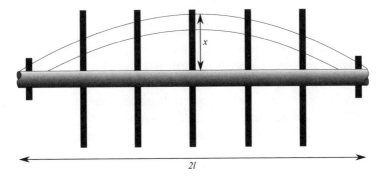

Imagine a section of pipeline 200 m long (so that $l = 100$ m). Seen from above, we look down on seven horizontal beams (*black bars*), on which the pipeline rests. The pipe can slide horizontally across the beams, except at the end points. I have made the end beams shorter, to suggest that the pipe at these points is fixed. If the air temperature heats up, the pipe will expand and buckle, as shown by the curved lines. The buckle distance, x, can be several meters.

drive-train friction, plus wheel friction on the road); thus, the total force, f, acting on the car is about 350 N. The energy expended in moving the car a distance L is $E = fL$.

This energy comes from the fuel. The total energy extracted from a tank of gasoline is $E = \epsilon DV$, where D is the fuel's volumetric energy density (32 MJ/L for gasoline), V is tank capacity, say 50 L, and ϵ is fuel-to-wheel efficiency, which is about 15% for a standard car with an internal combustion engine.

The two expressions for energy must be equal, and so we obtain the following rough estimate for distance traveled on a tank of gas:

$$L = \frac{\epsilon DV}{f} . \tag{A.9}$$

Substituting numbers gives a distance of 680 km (see chapter 5). The energy density of compressed natural gas is about 10 MJ/L, so the same car would travel only 210 km on a tank of compressed natural gas. As pointed out in chapter 5, this is why CNG-powered vehicles are big—they need big tanks.

The average power provided by the tank of gasoline is fv, and so the power generated by our engine during the 680-km drive is $P = fv/\epsilon$, where ϵ is now engine efficiency (say 25%). This works out at about 47 hp—a reasonable estimate for such a rough calculation.

Satellite Power

Consider a satellite in geosynchronous orbit, so that it stays in the same part of the sky as seen from an observer directly below it on the earth's surface. This satellite is at an altitude (let us call it A) of 35,600 km. It receives sunlight and converts the light energy into microwave energy, which it then transmits to a receiver on earth. Here, I will estimate the size of the satellite antenna, the size of the receiver on earth, and the power that such a satellite generator system could provide. The power transmitted down to earth as microwaves is

$$P_T = \frac{1}{4} \pi d^2 p_s \epsilon. \tag{A.10}$$

In words: the transmitted power equals the receiver antenna area multiplied by the power density of light from the sun (yielding the total solar power received), and multiplied by the efficiency of the satellite converting and retransmitting this power as microwaves. I am assuming that the receiver antenna is circular with diameter d. At the earth's surface this power is intercepted by a rectenna of diameter D, so that the received power is

$$P_R = \frac{1}{4} \pi D^2 p. \tag{A.11}$$

Let us say that the received power is limited, for safety reasons, to a maximum density of p. A reasonable value to assign for p is 10 W/m². If we further assume (this calculation is necessarily full of assumptions concerning the geometry of the system) for reasons of economy that the satellite antenna receiving sunlight is also used to transmit microwaves to earth, then the angular width of the transmitted beam is $\theta \approx 2\lambda/d$, where λ is the microwave wavelength. Let us take λ to be 10 cm, corresponding to a frequency of 3 GHz; this frequency can travel through the atmosphere with very little attenuation. To avoid wasting power, we fix the rectenna diameter to be the microwave beam width: $D = \theta A = 2\lambda A/d$. Neglecting sidelobe losses (reasonable, because we anticipate that the beam width will turn out to be very small) then we can say, approximately,

$$P_R \approx P_T. \tag{A.12}$$

From equations (A.10)–(A.12) we can solve for antenna diameters:

$$d \approx \sqrt{2}\left(\frac{p\lambda^2 A^2}{\epsilon p_s}\right)^{1/4}, \; D \approx \sqrt{2}\left(\frac{\epsilon p_s \lambda^2 A^2}{p}\right)^{1/4} \tag{A.13}$$

and power delivered to the grid:

$$P \approx \frac{1}{2}\ \pi\eta\lambda A \sqrt{\epsilon p_s p},$$ (A.14)

where η is the efficiency with which the rectenna converts received microwave power into usable electrical power.

I will assume efficiencies of $\epsilon = 0.5$ and $\eta = 0.8$. The solar power density at the satellite antenna is $p_s = 1.35$ kW/m^2 (see the introduction), so I obtain order-of-magnitude estimates for antenna diameters ($d = 1,300$ m, $D = 5,400$ m) and for deliverable power from such a satellite generator ($P = 735$ MW). This power is comparable to that provided by existing power stations.

NOTES

1. Newton's Legacy

1. All physicists were men in those days. The other half of humanity was, with a few exceptions, discouraged from contributing its talents to the mathematical sciences until the first half of the twentieth century. For the most profound definition of energy, and of certain other physical properties that are conserved, we must thank a physicist from this other half—Emmy Noether.

2. *Ingenuity* and *engineer* derive from the same medieval Latin root word, *ingenium*, or the Latin root *ingeniosus*. See Denny (2007) and Lienhard (2000).

3. Energy must be imparted to the atoms that make up a molecule of a chemical compound to force them together in a particular configuration. In this sense chemical energy is potential energy by virtue of the positions of the constituent atoms.

4. Atoms of a material jiggle faster and faster as the material gets hotter and hotter. In this view, thermal energy (heat) is thus a kind of kinetic energy.

5. Nuclear energy is weirdly different from other forms because it can be concentrated into mass, according to Einstein's famous equation. A little mass converts into a lot of energy. The nuclear reactions that take place in the sun, in a nuclear explosion, or in a nuclear reactor convert mass into other forms of energy—mostly heat.

6. If the main aim of the construction is to prevent the flow of water rather than retain it, it is known as a dike or levee.

7. *J* stands for joules, a unit of energy named for a nineteenth-century English physicist who was so keen on his subject that he reputedly took time out from his honeymoon to measure the temperature difference between the water at the top and the bottom of a scenic waterfall. (You see, there was a potential energy difference between the top and bottom and so, given the mechanical equivalent of heat, there should be a measurable difference in temperature.) See Cardwell (1989). *W* stands for watt, a standard unit of power named for the pragmatic Scottish engineer who did so much to develop steam engines and, consequently, the Industrial revolution. See Denny (2007), chap. 5.

8. At night and outside the tourist season. During peak tourist times the flow over the falls is maintained at a level that is much closer to the natural level of the Niagara River, by international agreement. See, e.g., Niagara (2010).

9. From the figures, we see that four typical lightning bolts would supply the total power needs of humanity—for a fraction of a millisecond. The numbers for lightning energy and power are from Hasbrouck (1996).

10. Figures for the energy conversion efficiency of different devices are widely available in the popular literature. See, e.g., Eden et al. (1983), Gipe (2009), Narisade and Schreuder (2004), and Parasiliti and Bertoldi (2003).

11. Heat itself weighs next to nothing, but the medium that contains it (such as air or steam) has weight.

12. We will see in chapter 3 that the "speed of electricity" is a slippery notion. The important point is that electrical power propagates down a line at a speed that is only a little less than the speed of light.

13. The process of redistributing electric power between automobile batteries and the power supplier, and the negotiation of rates both ways, is sometimes known as *carbitrage*.

14. A nontechnical overview of capacitors in electric cars can be found in Economist (2008a).

15. An Eveready Battery data sheet claims that one of this company's zinc-carbon batteries retains 65%–80% of its capacity after four years' storage. See Eveready (2001).

16. The battery industry refers to the positive electrode as the *cathode* and the negative electrode as the *anode*. Most people label the electrodes the other way around, with the cathode being negative. The reason for this difference—which causes confusion—is that *cathode* refers to the terminal toward which positive charge flows; inside the battery (the perspective of battery manufacturers), this is the positive terminal, whereas outside the battery, it is the negative terminal.

17. For more on BESS, see, e.g., Conway (2003) or the BESS website. For batteries in general, see Linden and Reddy (2002).

18. Usually these are *Francis turbines*, which were invented in mid-nineteenth-century America. The fact that such a venerable design is standard equipment 150 years later is testimony to its good design. These turbines are very efficient, as well as reversible. I will have more to say about Francis turbines in chapter 6.

19. The Brennan torpedo, an innovative late-nineteenth-century device, was powered by compressed air. See Denny (2011). In Victorian offices and factories, messages would be placed in tubes, which then fit inside extensive pipelines and were powered around a building by compressed air. The original "airmail," I guess.

20. A modern high-speed flywheel is placed inside a strong case for safety. The rotor is typically made of titanium and/or carbon fiber for strength (the centrifugal forces are huge) and is supported by magnetic bearings, which minimize axle friction.

21. Much has been written about energy storage, particularly in the context of smoothing the power supply (in the case of wind farms) and smoothing the grid power supply. For more details on this subject in general, the following books provide accessible accounts: Baxter (2006), Dell and Rand (2004), Huggins (2010), and Ter-Gazarian (1994). For a technical account of CAES in the context of sustainable energy, see Pickard, Hansing, and Shen (2009). For flywheels, see in particular Castelvecchi (2007) and Genta (1985).

22. Place two electrodes in a glass of water and pass an electric current through the liquid. The water will split up into its constituent elements: oxygen bubbles are formed at the anode and hydrogen bubbles at the cathode.

2. What All the World Wants

1. Boulton made similar statements many times over the years, in several different ways. This quote is from Restivo (2005).

2. The data for this section of text come principally from Cook (1971) and Mattick, Williams, and Allenby (2009).

3. For a discussion of energy density as a limiting factor in animal distribution, see Denny and McFadzean (2011).

4. The earliest draft "animal" was the slave. The donkey was more powerful and the mule more powerful still. Somewhat better was the ox—powerful and docile. Best of all was the horse, which generated between 5 and 10 times the power of a man and could be worked for more hours per day. Oxen and horses were the standard draft animals for most of history, until the machine age. See Carruthers and Rodriguez (1992). Bernal (1965) argues convincingly that the abundant use of slaves in the classical world delayed the development of man-made sources of power (such as waterwheels and windmills).

5. See Bernal (1965, p. 91) for the claim about power and Britannica (1998) for the more controversial assertion about waterwheels. The horse harness has a pretty good claim for this period, as it increased the pulling power of horses. Also, as we will see, there is an increasingly common view among historians that waterwheels were widely used before 500 CE.

6. Much has been written about the historical development and influence of waterwheels. For more details see Deane (1979), Denny (2007), James and Thorpe (1994), Landels (1978), Mason (1962), Stark (2005), and Usher (1988).

7. The flyball governor was the first widely used device for establishing feedback control in machinery—a hallmark of the Industrial Revolution and a significant step along the road toward modern control engineering. See, e.g., Buede (2009) and Denny (2002, 2007).

8. Instead of being limited in sail size by the height of the tower, the largest windmills were limited (to about 90 feet) by the length of pine shafts that held the sails.

9. Such *polder* land was surrounded by dikes, and water was pumped over the dikes so that the land (below sea level) would remain habitable.

10. The windmill has been analyzed and written about from many perspectives: engineering, economic, historical. To learn more about windmills, see Denny (2007), Hills (1996), James and Thorpe (1994), Tunis (1999), and Usher (1988). Usher reports that a typical turret windmill could generate 6–14 hp, whereas an older post windmill generated 2–8 hp. These figures compare favorably with 2–5 hp for a large (18-ft) overshot waterwheel.

11. The animation of a Newcomen engine at wikipedia.org/wiki/File:Newco men_atmospheric_engine_animation.gif demonstrates the operating principle quickly and effectively.

12. The steam engine and Cort's puddling process (a new means for making iron, invented by Henry Cort in 1783) are considered by Deane (1979) to be the two most important innovations that sustained economic growth during this period and made industrialization self-reinforcing and continuous.

13. An online animation conveys the action of a multiple-expansion, multi-cylinder steam engine: see wikipedia.org/wiki/File:Triple_expansion _engine_an imation.gif.

14. For early steam engine development, see, e.g., Buenstorf (2004), Deane (1965), Thurston (1878), and Usher (1988).

15. An unfortunate metaphor because in the early days, boilers (being subjected to high pressures) quite often did explode. Improved steel and improved production engineering reduced these disasters over the decades. Indeed, Watt was probably correct to oppose high-pressure engines in his day because the steel-making technology that made high-pressure engines safe was at least half a century in the future.

16. From 1,800 miles of track in 1840, European rail networks expanded to 45,000 miles by 1870. In the United States the expansion of rail was even more dramatic: from 3,300 miles in 1840 to 22,000 miles in 1860 and 192,000 miles by 1900. Railroads brought about a market revolution in the United States, changing it from a patchwork of local agricultural markets in 1800 to a national economy 100 years later. See Overy (2007).

17. Gustav de Laval in Sweden independently developed a variant turbine at about the same time. The concept of turbines was known much earlier, and the idea was sufficiently well regarded that Matthew Boulton worried that turbines might challenge the supremacy of his steam engines. Watt assured him that the slower engine would remain in the ascendancy: "Without God making it possible for things to move at a thousand feet a second [the turbine] cannot do much harm" (Mason 1962, p. 508).

18. To learn more about steam turbines and the men who developed them, see,

e.g., McNeil (1990), Rosen (2010), and Usher (1988). Wikipedia also has a useful entry on the subject.

19. For the characteristics of coal, see, for example, the article by Mancuso and Seavoy (1981). A very readable nontechnical account, with much on the early history of coal mining, is that of Freese (2003). The adverse effects on society of burning coal on an industrial scale are made very clear in Davies (2006).

20. Smelting iron with coal does not work: impurities in the coal contaminate the iron. The solution, found during the eighteenth century, was to first bake the coal to drive off volatile components, forming coke. Coke was then used to smelt iron.

21. This use of "damp" comes from the German *dampf*, meaning "fog" or "vapor."

22. This bald statement will annoy historians of technology: it is both too precise and too vague. Too precise because the internal combustion engine was developed over several decades, from about 1860, by people in France, Italy, Belgium, the United States, England, Austria, and Switzerland as well as Germany. Too vague because an internal combustion engine comes in many forms, including rocket engines. I chose 1862 because that is the year in which Nikolaus Otto first built and sold his engine; and Germany because Otto, along with fellow Germans Karl Benz and Gottlieb Daimler, figured prominently in the development of the automobile. You might choose to delay the invention of modern automobile engines to the 1870s, when in-cylinder compression became widespread.

23. To learn more about the historical development of automobiles, see, e.g., Berger (2001) and Flink (1990).

24. One indication of the influence of the automobile industry upon economies in the early twentieth century is the admiration that Adolph Hitler expressed for Henry Ford. This admiration was not only because of Ford's outspoken anti-Semitism but also because of his ideas for mass production. The car culture that evolved in 1930s Germany under Hitler led (along with rearmament) to a substantial German economic recovery following the hyperinflation of the Depression.

l25. Imagine that there were 43,000 deaths per year in the United States caused by terrorism. What would our response be like? Our attitudes to perceived risk depend very much on the cause; this is a drum I will beat again in this book, more than once.

26. The evolution of oil as a major industry and a source of transport fuel is well explained in, e.g. Maugeri (2006) and in Sampson (1975). Also useful is a U.S. government publication, USGAO (2005), available online. There are many sources of data for current energy demands and supplies. Each year the International Energy Agency publishes *Key World Energy Statistics*; see the agency's website at www.iea.org. Similarly the U.S. Energy Information Administration publishes an *Annual Energy Review*. Most of the data for this section comes from these publications.

27. Much of this section comes from Rajey (1996), Speight (2007), Tiratsoo (1983), and Tussing and Tippee (1995). Also, the natural gas industry maintains an informative website; see www.naturalgas.org.

28. The large U.S. deposit is the Haynesville shale, a 150-million-year-old black mudstone. In the days of low natural gas prices it made no commercial sense to extract from this shale deposit; now it does.

29. See the following resources for more on the historical development of hydropower, for information on dam construction physics and engineering, and for the figures for installed capacity: Denny (2010), Jackson (2005), Khagram (2004), Solomon (2010), and the useful websites maintained by the U.S. Energy Information Administration (international statistics database), the U.S. Department of Energy ("History of Hydropower") and the U.S. Bureau of Reclamation ("The History of Hydropower Development in the United States").

30. Prior to the detonation of the two bombs over Japan, a test explosion took place in the desert outside of Alamogordo, New Mexico.

31. For Chernobyl fatality statistics, see Yablokov, Nesterenko, and Nesterenko (2009); Greenpeace (2006); Chernobyl Forum (2006); and Hiserodt (2009).

32. A lot of ink has been spilled over all aspects of nuclear power development and reliability. See, e.g., Char and Csik (1987), Kaku and Trainer (1983), Morris (2007), and National Research Council (1992). There are also many online resources which provide insightful accounts, such as the U.S. Department of Energy's brief *History of Nuclear Energy*, at www.ne.doe.gov/pdfFiles/History.pdf; the International Energy Agency's *Key World Energy Statistics, 2010*, at www.iea.org/text base/nppdf/free/2010/key_stats_2010.pdf; and the World Nuclear Association website at http://world-nuclear.org/info/inf54.html.

3. The Vital Spark

1. We talk about the *transmission* of electrical power from generator site to local substation and the *distribution* of power from the substation to you, the customer. Transmission is usually over longer distances and with higher voltage than distribution.

2. An interesting account of the first 50 years of electric power is that of Hughes (1983).

3. "So, naturalists observe, a flea / Has smaller fleas that on him prey; / And these have smaller still to bite 'em, / And so proceed ad infinitum," according to the poet Jonathan Swift. In fact, physicists do not think that the world of matter constitutes such an infinite regress: electrons are likely truly elementary, made up of nothing more fundamental. The other components of atoms are protons and neutrons, which are made of elementary quarks. Electrons, protons, and neutrons will be discussed further in chapter 7.

4. The American Nobel laureate physicist and genius Richard Feynman expressed very well the importance of Maxwell's work: "From a long view of the history of mankind—seen from, say, ten thousand years from now, there can be little doubt that the most significant event of the 19th century will be judged as Maxwell's discovery of the laws of electrodynamics. The American Civil War will pale into provincial insignificance in comparison with this important scientific event of the same decade" (Feynman, Leighton, and Sands 1970).

5. "Joule's law" may sound like the name of a modern English actor, but in fact it refers to a nineteenth-century English physicist and brewer, James Joule, he whom the energy unit is named after.

6. See Riemersma et al. (1981).

7. The lines consist of twisted aluminum wires—twisted for reasons discussed in the text, and aluminum because it is lightweight—sometimes with an additional steel cable to add mechanical strength. If large currents flow, the conducting wire can be quite thick, in which case it is usually hollow; this reduces weight without increasing resistance (which is due to the skin effect).

8. An impressive YouTube video of electric arcing or flashover within a transmission line can be seen at www.youtube.com/watch?v=6GiIVze2Tac&feature= related.

9. Electricians in different countries use different conventions to color-code the three phases. Again, as with AC frequency, the reasons are largely historical.

10. The electrons themselves do not travel nearly so fast. For AC transmission they vibrate back and forth but do not move along the power line; for DC transmission they drift along at a few millimeters per hour. To understand this, I must again emphasize the difference between a wave and the particles that constitute it. Consider a wave of people at a baseball game. Each person stands up with arms raised (and a grin on his face) and then sits down. He does not move far from his seat, and yet the wave can travel around the baseball field in a few seconds.

11. A general rule for AC circuits: if generating power exceeds load, frequency increases; conversely, if load exceeds the power supplied, frequency decreases. Utilities take advantage of this fact to help them measure the imbalance between supply and demand.

12. See U.S.–Canada Power System Outage Task Force (2004).

13. For more on blackouts in general and the 2003 blackout in particular, see U.S.–Canada Power System Outage Task Force (2004), Lerner (2003), and Pérez-Peña and Lipton (2003).

14. That there are about 15,000 power stations of various sorts within the United States, delivering almost four trillion killowatt-hours of energy annually, should give you an idea of the size of these grids. Their power output is about 450 GW.

15. For the benefit of readers who know a little electrical theory, let me say some more about this mysterious reactive power (which is sometimes termed *phantom*

power). It is needed to power inductors and capacitors within a grid; these are the components which store energy and resist the flow of current, and which prevent voltage drop across a grid. They do not contribute to the end user's available power. Because reactive power does not travel far, it usually needs to be produced close to the place where it is needed. Reactive power—too much and too little—played a significant role in the 2003 blackout.

16. The Organisation for Economic Cooperation and Development consists generally of democratic free-market countries. Currently there are 34 members.

17. Figures for transmission and distribution losses are widely available. See, e.g., Chubu (2008, p. 38). For power pilfering in India, see Gregory (2006).

18. The high electrical capacitance of undersea cables increases the transmission losses of AC power.

19. In telecommunications and radar systems *waveguides* are sometimes employed to transmit electrical power; these take advantage of the wave nature of electromagnetic signals and trap them within confined regions. Waveguides work only when the waveguide width is of the same magnitude as the wavelength of the electromagnetic energy that it confines. They are large and bulky for microwaves and are widely used only for optical frequencies, where fiber-optic cables carry telecommunication signals over large distances with little loss.

20. Local governments and utilities have available a good deal of information for the public about the issues involved with underground vs. overhead transmission lines. See, e.g., Commission of the European Communities (2003), Edison (2009), PSC (2011), and SoCo (2009).

21. HVDC transmission is summarized in Casazza and Delea (2003). For a detailed technical account, see Arrillaga, Liu, and Watson (2007).

22. Thus, the U.S. Department of Energy YouTube video at www.youtube.com/watch?v=Qjyugl8Hncw&feature=related exudes an almost scaremongering concern about what will happen to the country if smart grids are not adopted. Smart grid capabilities are promoted in this and other clips—for example, by the World Economic Forum (www.youtube.com/watch?v=N8jqbKd8hVg) and Scientific American (www.youtube.com/watch?v=-8cM4WfZ_Wg&feature=related).

23. I recently heard a woman in a small-town store insist to her customers that smart meters were evil, in that they caused cancer (via emitted radiation) and permitted "them" to take over your life as well as increase your electricity bill. It seems that there were reams of data from California (where else?) to back up these findings. Perhaps such paranoia is one reason utilities and government organizations initiate education programs about smart grids and smart meters.

24. The goal of superconductor research is to create or discover a room-temperature superconductor. Intense effort over the last quarter century (represented by more than 100,000 research papers by 2001) has raised the maximum attainable superconducting temperature many times, and the current record is not likely to

last long. For accessible accounts of superconductivity, see Choi (2008) and Grant, Starr, and Overbye (2006). A technical appraisal of the frenetic research activity in the 1990s is provided by Buchanan (2001).

4. Old King Coal

1. For fossil fuel reserves and consumption statistics, see WCI (2005) and BP (2010). See also Simpson (2011).

2. Data are from Inman (2010), Termuehlen and Emsperger (2003), and Wright (2007).

3. Ironically, stockpiled coal was largely responsible for the failure of the long and bitter miners' strike in Great Britain in 1984–85.

4. See Bell, Davis, and Fletcher (2004).

5. For the U.S. study, see Abt Associates (2000).

6. A readable account of radioactivity resulting from coal burning can be found in Hvistendahl (2007). The Geological Survey analysis is in USGS (1997). I discuss this subject again in chapter 7.

7. The U.S. mortality figures are from "Coal Daily Fatality Report—Year End 2010," a summary table on the Department of Labor's Mine Safety and Health Administration website at www.msha.gov/stats/charts/coal2010yearend.asp.

8. For data on Chinese mines, see Gregory (2010), and the U.S. Mine Rescue Association website at www.usmra.com/chinatable.htm.

9. The greenhouse effect works as follows: Visible light passes through air quite easily (which is why we see the sun) but is absorbed by greenhouse gases. Thus, some of the sunlight that penetrates our atmosphere is soaked up by molecules of greenhouse gas, say CO_2. These molecules then re-radiate the absorbed energy at infrared wavelengths. Infrared radiation heats the atmosphere.

10. For greenhouse gas data in general, and atmospheric CO_2 in particular, see Boden, Morland, and Andres (2011); Keeling et al. (2009); and the websites of the National Oceanic and Atmospheric Administration (NOAA) and the Carbon Dioxide Information Analysis Center (CDIAC), Oak Ridge National Laboratory.

11. Countless, but fewer than formerly because of deforestation.

12. A discussion of atmospheric CO_2 and the 3% figure for anthropogenic contributions can be found at the U.S. Global Change Research Information Office website at, e.g., www.gcrio.org/ipcc/qa/05.html.

13. Burning coal uses up atmospheric oxygen as well as adding atmospheric carbon dioxide. Here at least, we do not need to worry. The 980 billion tons of coal will, when burned, deplete the air of only 0.2% of its oxygen. (Our atmosphere consists, at present, of 20.95% oxygen.)

14. A popularizer of energy concepts, Professor Smil has written a number of books and papers on the subject of future energy options. See, e.g., Smil (2010a).

15. See Smil (2010a).

16. Britain's chief scientific officer, David MacKay, has noted that, were fossil fuel power plants to be entirely replaced by plants fueled by renewable sources, they would need to take up half the area of his country, because of their relatively low power area density. See Kahya and Anderson (2011).

17. Smil seems to revel in generating such unwelcome facts about power generation and in exploding popular myths.

18. The only comprehensive references on mining are textbooks for students or mining engineers. The popular literature tends to concentrate on mining history, mine safety, or environmental issues. There are only a few accessible books that describe day-to-day mining operations and the machines that are used. Try Demont (2009), Dix (1988), or Lockard (1998).

19. Some of these monsters can be seen in an entertaining YouTube video, at www.youtube.com/watch?v=GF-c5Tn-Xhk&feature=related.

20. The Science Channel video at http://science.discovery.com/videos/how-do-they-do-it-coal-mining.html describes coal mining and processing very clearly.

21. Glider pilots and migrating birds know to use the thermals generated by power plant cooling towers to gain height for little expenditure of their own energy.

22. For the mercury contamination of fish, see Dean (2009).

23. The U.S. Energy Information Agency estimates the current cost of coal-generated electrical power as \$95 per MWh, and the cost of power generated by new technology as \$136 per MWh. See EIA (2010c).

24. An accessible and well-balanced overview of clean coal technology can be found at the National Energy Board of Canada website, www.neb.gc.ca/clf-nsi/rnr gynfmtn/nrgyrprt/lctrcty/clfrdpwrgnrtn2008/clfrdpwrgnrtnnrgybrf-eng.html.

25. There are over 100 gasification plants worldwide today, and this number is forecast to increase 5% annually. Another advance results from increased understanding of the burning process. Fluid dynamics and burn temperature affect the amount of nitrogen oxide pollutants that are generated. Thus, the quantity of these pollutants can be minimized by appropriate burner design and operation.

26. Some CCS occurs prior to combustion. Gasifying coal yields CO_2 as well as carbon monoxide and hydrogen. Removing CO_2 from the mixture leaves syngas; this removal process is part of CCS.

27. Underground storage of carbon dioxide is reported in, e.g., Kahya and Anderson (2011).

28. See, e.g., the following two YouTube videos, one from either side of the debate: www.youtube.com/watch?NR=1&v=ZmVDu_gIpc4, and www.youtube.com/watch?v=BgOEQLqPkuY&feature=related.

29. The World Bank initiative is discussed by Harvey (2011). For the EPA initiative, see Crooks (2011).

30. Apart from textbooks, most books about coal focus on history or on cur-

rent or past pollution. A good general history is that of Freese (2003). Two books that look to the future, on opposite sides of the debate, are Goodell (2006) and Mostrous, Gue, and Dittman (2010).

5. The Seven Sisters—Old and New

1. Oil fields in Texas, California, and other contiguous U.S. states figured prominently during the early years of oil extraction, but demand soon outstripped domestic supply, and so the United States turned abroad for its oil, as we saw in chapter 2.

2. I defy anyone to explain to me why it makes sense for the leaders of Iran to threaten to close the Straits of Hormuz (through which 30% of the world's oil—a lot of it Iranian—is shipped). They do so as I write these words, in January 2012. For a discussion on trade choke points, see Bernstein (2008, chap. 13).

3. The upper estimate of 25 years for recoverable oil supply is from "Oil: Crude and Petroleum Products Explained," Energy Information Administration website, www.eia.gov/energyexplained/index.cfm?page=oil_home#tab2. Many of the statistics in this section are obtained from this EIA website. We will see that this figure is considered optimistic by some experts.

4. See, e.g., Economist (2008b).

5. The original Seven Sisters are described by Anthony Sampson in a well-known book that delves into the shenanigans these oil companies indulged in to maintain their dominance; see Sampson (1975). The new Seven Sisters are an invention of the *Financial Times* newspaper. See Hoyos (2007).

6. The truck is moved to a desired location. There, it lifts a heavy weight to a height of about 10 feet and drops it to the ground repeatedly. Analysis of how the resulting thumps reflect off underground geological structures forms part of the spatial data that can be used to construct images of what is beneath the ground.

7. The new reservoirs are estimated to contain 700 million barrels of recoverable oil. Because they are under 7,000 feet of water, extraction will be a challenge, especially given the nearby BP Deepwater Horizon oil disaster. BP and Chevron have also made significant finds recently in the Gulf of Mexico.

8. An example is the video of "whipstock" drilling at www.youtube.com/watch?v=JGvIHdyPYG4.

9. See "Offshore Rig Day Rates," Rigzone website, www.rigzone.com/data/day rates for current rates.

10. A helpful video explaining the different recovery phases is at www.youtube .com/watch?v=Utmao53t8xo&feature=related. For a more detailed account, see Tzimas et al. (2005).

11. The figures are from Trench (2001) and refer to 1999. An example of geographical factors is the difficulty in the continental United States of transporting oil

by sea from coast to coast. For Indonesia or Chile, on the other hand, transportation by sea is much easier than by pipeline.

12. Crude is hot when it emerges from the ground, which is good news for pipeline transportation because hot oil is less viscous than cold oil and is therefore easier to pump. Pumps are placed every 20–100 miles to facilitate oil movement. Oil takes several days to travel through a long pipeline because the fluid speed is typically a few miles per hour. (See the TAPS calculations in the appendix for a worked example.) The pipes are kept clean by mechanical pipeline inspection gauges (called PIGs) that travel inside the pipes removing wax buildup. Monitoring for leaks reduces environmental damage and theft.

13. The numbers are from UN (2006) and from the U.S. Energy Information Administration. Only tankers below a certain size can pass through the Panama and Suez Canals. In the 1970s, because of political instability around the Suez Canal, tanker companies adopted alternative routes. Although these routes were longer and therefore more expensive, without the canal size constraint the tankers could become bigger, so that economies of scale helped offset the increased route cost.

14. This incident was reported in the *Juneau Empire* on October 5, 2001; see Clark (2001).

15. An oil pipeline explosion in Nigeria that killed 200 people in 2006 was linked to oil theft by local villagers (AP 2006); in 2009 Nigerian militants blew up another pipeline (CNN 2009).

16. See Stec and Baraj (2007). For a report on the Libyan pipeline shutdown, see Blas (2011).

17. For example, see the U.S. Energy Information Administration's map at www.eia.gov/todayinenergy/detail.cfm?id=330.

18. In addition to the recent Iranian threat to close the Strait of Hormuz, the political temperature has been raised in the past over the nationalization of the Suez and Panama Canals. At the time of this writing, the Reuters news agency was reporting that a U.S. oil tanker had been hijacked by Somali pirates (Reuters 2011).

19. Data from the U.S. Energy Information Administration website, www.eia .gov/energy_in_brief/foreign_oil_dependence.cfm.

20. The subscript numbers indicate how many atoms of each element are in a molecule.

21. The *New York Times* has quoted the International Energy Agency as saying that peak oil was in 2006 (Rudolf 2010).

22. A very good general reference about the coming decline in oil is that of Roberts (2004).

23. Jeroen van der Veer quoted in *Calgary Herald* (2008); Caruso (2005).

24. David Greene, questioned during an AAAS webcast (AAAS 2012).

25. One possible indicator of oil scarcity is the efficiency with which we use it.

Efficiency can be measured by *oil intensity*, which is the amount of oil required to generate a dollar of GDP. In the United States and other Western countries, oil intensity has halved since the 1980s. See, e.g., Staniford (2005).

26. Alan MacFadzean, personal communication, March 2012.

27. Peter Ward, personal communications, March 2012.

28. A YouTube video of the Chevron Endeavor fire can be seen at www.youtube .com/watch?v=28c3Yrq5Qlc.

29. See Donnelly (2005), or Leeb (2009).

30. For more on shale oil, see, e.g., Bunger, Crawford, and Johnson (2004); and Johnson, Mercier, and Brownfield (2011). See also the American Petroleum Institute factsheet, "U.S. Oil Shale: Our Energy Resource, Our Energy Security, Our Choice," at www.api.org/oil-and-natural-gas-overview/exploration-and-product ion/oil-shale/energy-resource.aspx.

31. The publicity surrounding the Athabasca tar sands has meant that a great deal of information is available in the media and online. For technical information and an informative video from the petroleum industry perspective, see Canadian Association of Petroleum Producers (2010). A view of the health and environmental damage from tar sands opponents is given in the Canadian Broadcasting Company documentary "Tipping point: The age of the oil sands," www.cbc.ca/docu mentaries/natureofthings/2011/tippingpoint. See also Haggett (2009) and Timoney and Lee (2009).

32. MIT (2011). Urbina (2011) is an interesting series of articles in the *New York Times* about drilling for natural gas and regulating the industry. For a concise, enlightening article on the extraction and transportation of natural gas, see Pynn (2012). On Canada's west coast, the prospect of pipelining natural gas to the coast for export to Asia is raising both eyebrows and hackles; see Hamilton (2012).

33. As I write these words (March 5, 2012) a freak automobile accident in Illinois has caused oil spillage and the shutting down of the Enbridge pipeline, which supplies the United States with Canadian oil. See Fox (2012).

34. The National Commission final report is available online at www.oilspill commission.gov/final-report. Some of the more insightful press investigations include Flynn (2012), Memoli and Nicholas (2010) and Urbina (2011).

35. Raw data is from the *CIA World Factbook*, www.cia.gov/library/publica tions/the-world-factbook/index.html.

6. Water, Water, Everywhere

1. For potential hydro capacity see, e.g., Ray (2010). National data were obtained from the International Energy Agency, "Statistics and Balances," www.iea .org/stats.

2. Boyle (2004) is a good general reference on all things hydro. The economics of hydropower, as well as the pros and cons, are discussed in detail in Edwards (2003).

3. There are also a lot of micro hydro plants in the first world. Many hobbyists and environmentalists with access to running water like to construct waterwheels or small turbines to generate some electricity. I wrote a physics paper about waterwheel efficiency (Denny 2004), and such is the interest in the subject, I receive more requests for this paper than for any other I have written.

4. For dam operating costs, see (e.g.) the U.S. Bureau of Reclamation, "Reclamation-Wide Power Profile," available at www.usbr.gov/power/data/recl-wid.pdf.

5. For more details on dam design, see Denny (2010).

6. In addition to the hydropower references cited earlier, see also the excellent online PowerPoint presentation of Prof. Stephen Lawrence of the University of Colorado, at www.slidefinder.net/h/hydropower_professor_stephen_lawrence_l eeds/hydropower/23940232.

7. In fact the main (lunar) component of the tides is not quite twice daily: the period between high tides is 12 hours and 25 minutes. The period of the solar component is 12 hours. The interaction of these two components gives rise to unusually high and unusually low tides (*spring* and *neap* tides, respectively) every two weeks. The height of the tides in different parts of the world depends very much upon local shoreline topography and coastal bathymetry. At a given location the tide height varies with meteorological conditions.

8. Incoming tidewater overtops the barrage or passes through sluices. As the tide recedes (ebbs), this water is retained, so hydraulic head is built up, reaching a maximum at low tide. Electrical power is then generated via turbines, as for a conventional impoundment dam hydropower station.

9. For technical details on tidal power see, e.g., Clark (2007), Greenberg (1979), and Karsten et al. (2008). The badly edited book by Charlier and Finkl (2009) is also worth a look. There are also many informative nontechnical online sources, such as Wikipedia.

10. The turbines that are being used to extract power from tidal streams (such as the *straflo*—from "straight flow"—turbine) are different from the turbines that we employ in conventional hydro plants because the flow conditions are different.

11. Perhaps in the far future there will be hundreds of square miles of ocean bed given over to millions of tidal stream turbines, tapping into well-known deep ocean currents, but it won't happen before the oil runs out.

7. Too Cheap to Meter

1. Nuclear energy is the only energy source on earth that does not originate from within the solar system. Astrophysicists tell us that the elements which we use

for nuclear fission (indeed, all elements heavier than helium) were created in ancient stars that later exploded. Our solar system was formed from the debris. While the chemical energy of atoms and molecules results from reactions that occur within the solar system, the energy of their nuclei comes from some distant, dead star. When fissionable material such as uranium and plutonium releases energy, it is the energy from this dead star.

2. In fact, protons and neutrons are made up of different types of more fundamental particles—quarks—and so are not truly elementary. It is believed that quarks and electrons *are* truly elementary.

3. For comparison, you may recall that the energy density of hydrogen is greater than that of fossil fuels, at 38 kWh/kg.

4. Gas centrifuge technology is politically contentious because it can be used to enrich uranium to weapons grade. The concerns at the time of this writing about alleged Iranian nuclear weapons programs center on their enrichment capabilities. Mohamed ElBaradei, the Nobel Peace Prize winner and former director general of the International Atomic Energy Agency, has proposed that nuclear nations provide nonnuclear countries with access to enriched uranium fuel without providing enrichment technology, so that the nonnuclear countries may build nuclear power plants without being able to develop nuclear weapons.

5. See, e.g., Bedi (2011), Bullen (2005), Cocks (2009), Fetter (2009), Lightfoot et al. (2006), Nersesian (2010), Pendergast (1991), and Sovacool (2011). See also the World Nuclear Association, "Supply of Uranium," www.world-nuclear.org/info/inf75.html, for an interesting discussion about natural resource estimation in general, and nuclear fuel resources in particular.

6. Many books and online accounts explain nuclear technology; the problem is not finding information but weighing it. For PWR design, see, e.g., NRC (2010), Peters et al. (2005), the Wikipedia website, and the YouTube video at www.youtube.com/watch?v=MSFgmLW1Crw. These sources are a small subset of all that are available.

7. As a physicist, I cannot avoid noting that all four of the fundamental forces of nature (strong and weak nuclear forces, electromagnetic force and gravity) are involved in the design of reactors with this basic failsafe mechanism.

One important reason that PWRs are considered very safe involves another failsafe feature: should coolant temperature increase beyond the operational range, the natural consequence of the design is to reduce reactor core temperature. (Heating causes the water to expand, reducing its ability to moderate. Consequently, neutrons are not slowed down and absorbed, and so the nuclear reaction slows and cools.) For Chernobyl-type reactors the opposite occurs: an increased coolant temperature leads to increased core temperature which further increases coolant temperature, in a runaway reaction that, as we know, ends badly. More on this matter later.

8. "Heavy water" is one of the nuclear technology phrases that has made its way into popular culture. In World War II a commando operation in Norway destroyed a German heavy water facility that was part of a Nazi atomic bomb research project. The raid later became the subject of a Hollywood film, *The Heroes of Telemark*.

9. One molecule in 6,400 of naturally occurring water is deuterium, and so each of us has a gram or so of deuterium in our bodies. Drunk in very large amounts it can be harmful because its diffusion properties are a little different from those of light water. It is not radioactive.

10. Despite circumventing enrichment, CANDU reactors probably helped provide the skills necessary for the Indian government to develop nuclear weapons. (The country has CANDU reactors and a number of other reactors of derivative design.) After India had developed a nuclear bomb in 1974, Canada ceased all exchanges of nuclear technology with that country.

11. See, e.g., Hall (2011).

12. For more on fast breeder nuclear reactors, see Karam (2006); for the changing Japanese policy, see Daly (2012). As with all aspects of nuclear energy, there are any number of online articles explaining FBR technology (e.g., Wikipedia).

13. In the late 1980s, hopes were raised briefly by two reputable researchers who claimed to have found an entirely new and affordable method for releasing energy through fusion at low temperatures. Such *cold fusion* could be achieved on a very small scale—on your kitchen table—and at low cost. The idea received a great deal of media attention (it might have solved the energy problem overnight) and a lot of scientific scrutiny. Unfortunately, the initial results could not be reliably reproduced, and although cold fusion is still being researched today, it is not widely considered to be viable, not least because there is no theoretical understanding of what mechanism might be responsible for the claimed results. It's a pity: the world would be a different place today had cold fusion proved to be more than a flash in the pan.

14. For readable accounts of fusion technology and the current state of play, see, e.g., Hickman (2011), Svoboda (2011), and the World Nuclear Association article "Nuclear Fusion Power," www.world-nuclear.org/info/inf66.html.

15. For example, the British advanced gas-cooled reactor (AGR) uses carbon dioxide as a coolant. AGRs are a second-generation design based on the earlier Magnox reactors, which proved the technology but were uneconomical for commercial use.

16. A whimsical note on the very serious subject of nuclear proliferation: While researching the subject online, I found an eBay ad announcing "fantastic prices on Pu-239." Happily, the ad was for a movie, not the nuclear fuel.

17. In 2009 *Scientific American* reported that "the U.S.—and the world—is gearing up to build a potentially massive fleet of new nuclear reactors, in part to fight climate change." See Scientific American (2011) for this quote and for several informative articles on the subject of nuclear power.

18. See Economist (2011).

19. See, e.g., AFP (2012), Kim (2011), and Paul-Choudhury (2011).

20. We will see later on that direct comparisons of radiation levels can be misleading.

21. About the same number of people die every year across the world from allergic reaction to peanuts.

22. See, e.g., Caracappa (2011) and IAEA (2011).

23. I recall from years ago a sad letter to the *Times* of London from a family doctor in the north of England who was dying from a rare form of cancer at the age of 46. He lived and worked close to a nuclear power plant and blamed his condition on radiation received from this plant. But was he right? We will never know. Even if he was right, how could the plant be held responsible? There was no smoking gun to say whether this particular man got his fatal dose from the nuclear power plant or from natural sources.

24. Statistics on automobile fatalities were obtained from the National Highway Traffic Safety Administration website at www-fars.nhtsa.dot.gov/Main/index.aspx.

25. See Hvistendahl (2007) and Yulsman (2008). The radioactive nature of fly ash from coal-fired power plants is discussed in a USGS fact sheet (USGS 1997), which concludes that the natural radioactivity of fly ash should not be a source of alarm, but that the fact "provides a useful geologic perspective for addressing societal concerns regarding possible radiation and radon hazards."

26. Total costs include design and construction of a plant as well as operating and maintenance costs. They also include insurance, financing, lifetime effects such as upgrades, ecological and environmental measures, and so forth. In chapter 8, I summarize the economic costs of different technologies as determined by people more versed in such matters than I am.

27. For screening curves in general, see Masters (2004). The data in this section are from Green, Hu, and Vasilakos (2011).

8. Here Comes the Sun

1. Heating and cooling of the earth's surface by day and night, and differential heating of the land and ocean, causes air to expand and rise or contract and fall, giving rise to lateral air movement.

2. The advantages and disadvantages of vertical-axis wind turbines reflect the pros and cons of earlier vertical-shaft windmills, which we met in chapter 2. To a wind farm engineer, the important distinction is between efficient VAWT or HAWT turbines (such as propellers) that generate aerodynamic lift and inefficient VAWT turbines (such as anemometers) that rely solely on aerodynamic drag.

3. This is a rough rule of thumb; in any single case the increase of air speed with altitude can be a lot more or less than one-seventh. For instance, over ice the wind

speed increases as height raised to the power 0.07 because there is little friction between the wind and the smooth surface of the ice. Over woodlands, the exponent is 0.43; over prairie it is 0.18.

4. Some relatively minor environmental issues are associated with wind-power plants. They can be noisy (especially the smaller turbines, which are located closer to populations than are large wind farms), and they also kill birds that happen to fly through the swept area. In addition, air turbulence resulting from the action of turbine blades causes a significant reduction in the daytime-to-nighttime temperature fluctuation on a wind farm site. This last factor may be the most significant ecological effect when wind farms become more widespread.

5. Two very informative publications from the U.S. Department of Energy shed a great deal of light on the technical aspects of wind farming, particularly the integration of large wind farm power with grids, and on government attitudes toward wind farms and the promise they hold; see DOE (2008) and (2011). For wind data see, e.g., Garvine and Kempton (2008). For micro-turbine operation and domestic wind farming, see Gipe (2009). Argonne National Laboratory has a useful educational website on wind energy at windeis.anl.gov/guide/basics/index.cfm.

6. For those readers who know some physics, I note here that the photovoltaic effect is similar to the photoelectric effect (for the correct explanation of which Einstein was awarded his Nobel Prize—not for relativity), except that electrons are not ejected entirely from the medium (semiconductors, in the case of solar panels) by photons of sunlight but are instead bumped up from valence to conduction band energies within the medium.

7. In Europe electrical power from photovoltaics costs €0.22 per kWh, compared with €0.10 per kWh from natural gas, €0.09 per kWh from onshore wind, and €0.05 per kWh from nuclear.

8. Quoted in Cartlidge (2011). For more on various aspects of solar power, see Connolly (2012), Evans (2012), Hargreaves (2011), Smil (2005), and Swanson (2009). Boxwell (2011) explains the practicalities of domestic solar power.

9. Mankins (2002) provides an overview of this type of satellite power generator.

10. As always for this type of estimation, the numbers could easily be out by a factor of two.

11. A recent, very comprehensive international study of the potential for space solar power is available in print and online: see Mankins (2011) and the associated International Space Solar Power Symposium lecture summarizing this report at www.nss.org/resources/library/videos/SSPmankins.html. I am grateful to Dr. Clayton Lewis for providing this reference.

12. In joining the long list of writers who make predictions about our long-term energy future, I am taking a risk and inviting derision. For example, Smil is scathing about the "laughable predictions" of energy forecasters. See Smil (2005).

13. Access to geothermal energy is severely limited by the very few sites avail-

able; biofuel takes away agricultural land we will need for growing food and is thus self-limiting.

14. As much as 5% of U.S. energy is consumed by machines in standby mode. Perhaps this would be a good place to start economizing.

Afterword

1. Hoffert, quoted in Parfit (2005); MacFadzean, personal communication, March 2, 2012. There are a myriad of editorial warnings of catastrophe from both sides of the energy debate (for example, economic catastrophe if we sign up for climate change emission reductions, and environmental catastrophe if we don't). A succinct example is that of Simpson (2011).

2. The relationship between GDP and energy consumption is more complicated than the choices presented here suggest; see, for example, figure 5.7.

Appendix

1. The Tennessee Valley Authority maintains an extensive website containing many educational and readable accounts of all aspects of energy and power generation and is well worth a browse for readers who want to delve a little deeper into our subject.

BIBLIOGRAPHY

AAAS (American Association for the Advancement of Science). 2012. "Peak oil: Is the well running dry?" Webcast. Feb. 2. http://news.sciencemag.org/scienceno w/2012/02/live-chat-peak-oilis-the-well-ru.html.

Abt Associates. 2000. *The Particulate-Related Health Benefits of Reducing Power Plant Emissions.* Bethesda, MD, October. www.abtassociates.com/reports/par ticulate-related.pdf.

AFP (Agence France-Presse). 2012. "Japanese tsunami death toll at 19,300 10 months on." *Vancouver Sun,* Jan. 10.

AP (Associated Press). 2006. "Nigeria ups security after deadly pipeline blast." Posted online May 13. www.msnbc.msn.com/id/12754908/ns/world_news-af rica/t/nigeria-ups-security-after-deadly-pipeline-blast.

Arrillaga, J., Y. H. Liu, and N. R. Watson. 2007. *Flexible Power Transmission—The HVDC Options.* Chichester, UK: John Wiley.

Baxter, R. 2006. *Energy Storage: A Nontechnical Guide.* Tulsa, OK: PennWell Corp.

Bedi, R. 2011. "Largest uranium reserves found in India." *Telegraph,* July 19. www .telegraph.co.uk/news/worldnews/asia/india/8647745/Largest-uranium-reser ves-found-in-India.html.

Bell, M. L., D. L. Davis, and T. Fletcher. 2004. "A retrospective assessment of mortality from the London smog episode of 1952: The role of influenza and pollution." *Environmental Perspectives* 112:6–8.

Berger, M. L. 2001. *The Automobile in American History and Culture.* Westport, CT: Greenwood.

Bernal, J. D. 1965. *Science in History.* London: Penguin.

Bernstein, W. 2008. *A Splendid Exchange: How Trade Shaped the World.* London: Atlantic Books.

Blas, J. 2011. "The politics of oil: Wells of anxiety." *Financial Times,* March 29. www.ft.com/intl/cms/s/0/73d9207e-5a4b-11e0-8367-00144feab49a.html#axz z1oNOJfDZ8.

Boden, T. A., G. Morland, and R. J. Andres. 2011. *Global, Regional, and National Fossil-Fuel CO_2 Emissions.* Carbon Dioxide Information Analysis Center, Oak

Ridge National Laboratory, U.S. Department of Energy, Oak Ridge, TN. DOI: 10.3334/CDIAC/00001_V2011.

Boxwell, M. 2011. *Solar Electricity Handbook*. Ryton on Dunsmore, UK: Code Green.

Boyle, G. 2004. *Renewable Energy*. Oxford: Oxford University Press. Chap. 5.

BP (British Petroleum). 2010. *Quantifying Energy: BP Statistical Review of World Energy, June 2010*. www.bp.com/statisticalreview.

———. 2011. *Statistical Review of World Energy 2011*. www.bp.com/sectionbodyco py.do?categoryId=7500&contentId=7068481.

Brekke, P. 2004. "Space weather effects on SOHO and its space weather warning capabilities." In *Effects of Space Weather on Technology Infrastructure*, ed. I. A. Daglis, 109. NATO Science Series 176.

Britannica. 1998. *Encyclopaedia Britannica*. CD 98 Standard Edition. S.v. "Water wheels."

Buchanan, M. 2001. "Mind the pseudogap." *Nature* 409:8–11.

Buede, D. M. 2009. *The Engineering Design of Systems*. Hoboken, NJ: John Wiley and Sons.

Buenstorf, G. 2004. *The Economics of Energy and the Production Process: An Evolutionary Approach*. Cheltenham, UK: Edward Elgar Publishing.

Bullen, D. B. 2005. "Nuclear fuel reserves." In *Kirk-Othmer Encyclopedia of Chemical Technology*. Published online Oct. 14. DOI: 10.1002/0471238961.14210312 02211212.a01.pub2.

Bunger, J. W., P. M. Crawford, and H. R. Johnson. 2004. "Is oil-shale America's answer to peak-oil challenge?" *Oil and Gas Journal*, Aug. 9. http://fossil.energy .gov/programs/reserves/publications/Pubs-NPR/40010-373.pdf.

Calgary Herald. 2008. "Shell predicts energy shortage by 2015." *Calgary Herald*, Jan. 26. Online article at www.canada.com/calgaryherald/news/calgarybusiness/stor y.html?id=09669612-9de3-4278-b882-d5c9a16c89bc, retrieved March 5, 2012.

Canadian Association of Petroleum Producers. 2010. "Canada's oil sands: Come see for yourself." Online video. January. www.capp.ca/canadaIndustry/oilSands/ Dialogue-Resources/oil-sands-videos/Pages/Oil-Sands-Tour.aspx#1Fp2T4bL WdRL.

Caracappa, P. F. 2011. "Fukushima accident: Radioactive releases and potential dose consequences." Presentation at ANS Annual Meeting, June 28. www.ans .org/misc/FukushimaSpecialSession-Caracappa.pdf.

Cardwell, D. S. L. 1989. *James Joule: A Biography*. Manchester, UK: Manchester University Press.

Carruthers, I., and M. Rodriguez. 1992. *Tools for Agriculture: A Guide to Appropriate Equipment for Smallholder Farmers*. Rugby, UK: ITDG Publishing.

Cartlidge, E. 2011. "Boom times ahead for solar power." *Physics World* 24:10–11.

Caruso, G. 2005. "When will world oil production peak?" Presentation at 10th

Annual Asia Oil and Gas Conference, Kuala Lumpur, Malaysia, June 13. http://industrializedcyclist.com/EIAspeakoil4cast.pdf.

Casazza, J., and F. Delea. 2003. *Understanding Electric Power Systems: An Overview of the Technology and the Marketplace.* Hoboken, NJ: John Wiley and Sons.

Castelvecchi, D. 2007. "Spinning into control." *Science News* 171:312–13.

Char, N. L., and B. J. Csik. 1987. "Nuclear power development: History and outlook." *IAEA Bulletin*, 19–25.

Charlier, R. H., and C. W. Finkl. 2009. *Ocean Energy: Tide and Tidal Power.* Berlin: Springer.

Chernobyl Forum. 2006. *Chernobyl's Legacy: Health, Environmental and Socio-Economic Impacts.* Vienna, Austria: International Atomic Energy Agency.

Choi, C. Q. 2008. "Iron Exposed as High-Temperature Superconductor." *Scientific American*, June.

Chubu. 2008. *Chubu Electrical Power Group CSR Report.* September. www.chuden.co.jp/english/resource/csrreport2008.pdf.

Clark, M. 2001. "Pipeline pierced by bullet." *Juneau Empire*, Oct. 5. http://juneauempire.com/stories/100501/sta_pipeline.shtml.

Clark, R. H. 2007. *Elements of Tidal-Electric Engineering.* Hoboken, NJ: John Wiley and Sons.

CNN (Cable Network News). 2009. "Nigerian militants claim pipeline blast, tanker crew's seizure." Posted July 6. http://edition.cnn.com/2009/WORLD/africa/07/06/niger.militants/index.html.

Cocks, F. H. 2009. *Energy Demand and Climate Change.* Weinheim, Germany: Wiley-VCH.

Commission of the European Communities. 2003. "Background paper: Undergrounding of electricity lines in Europe." Brussels, Dec. 10. http://ec.europa.eu/energy/gas_electricity/studies/doc/electricity/2003_12_undergrounding.pdf.

Connolly, K. 2012. "Germany to cut solar power subsidies." *Guardian*, March 2. www.guardian.co.uk/world/2012/mar/02/germany-cuts-solar-power-subsidies.

Conway, E. 2003. "World's biggest battery switched on in Alaska." *Telegraph*, Aug. 28. www.telegraph.co.uk/technology/3312118/Worlds-biggest-battery-switched-on-in-Alaska.html.

Cook, E. 1971. "The flow of energy in an industrialized society." *Scientific American* 224:134–44.

Crooks, E. 2011. "U.S. coal-fired power stations under threat." *Financial Times*, Aug. 28. www.ft.com/intl/cms/s/0/0cc83d4e-cf0c-11e0-86c5-00144feabdc0.html#axzz1fWl2kXOu.

Daly, J. C. K. 2012. "Some dreams die hard: Japan's fast breeder reactor program." StockMarketsReview.com, March 6. www.stockmarketsreview.com/extras/some_dreams_die_hard_japan039s_fast_breeder_reactor_program_288465.

Davies, A. 2006. *The Pit Brow Women of the Wigan Coalfield*. London: History Press.

Dean, C. 2009. "Mercury found in every fish tested, scientists say." *New York Times*, Aug. 19. www.nytimes.com/2009/08/20/science/earth/20brfs-mercuryfound _brF.html?_r=1&em.

Deane, P. 1965. *The First Industrial Revolution*. Cambridge: Cambridge University Press.

Dell, R. M., and D. A. J. Rand. 2004. *Clean Energy*. Cambridge: Royal Society of Chemistry.

Demont, J. 2009. *Coal Black Heart: The Story of Coal and Lives It Ruled*. Toronto: Doubleday Canada.

Denny, M. 2002. "Watt steam governor stability." *European Journal of Physics* 23: 339–51.

———. 2004. "The efficiency of overshot and undershot waterwheels." *European Journal of Physics* 25:193–202.

———. 2007. *Ingenium: Five Machines That Changed the World*. Baltimore: Johns Hopkins University Press.

———. 2010. *Super Structures: The Science of Bridges, Buildings, Dams, and Other Feats of Engineering*. Baltimore: Johns Hopkins University Press.

———. 2011. "Depth control of the Brennan torpedo." *IEEE Control Systems Magazine* 31:66–73.

Denny, M., and A. McFadzean. 2011. *Engineering Animals: How Life Works*. Cambridge: Harvard University Press.

Dix, K. 1988. *What's a Coal Miner to Do? The Mechanization of Coal Mining*. Pittsburgh: University of Pittsburgh Press.

DOE (U.S. Department of Energy). 2005. *Peak Oil—The Tipping Point*. Energy Information Administration pamphlet. www.fossil.energy.gov/programs/re serves/npr/publications/Peak_Oil_-_the_Tipping_Point_final.pdf.

———. 2008. *20% Wind Energy by 2030*. July. DOE/GO-102008-2567.

———. 2011. *Strengthening America's Energy Security with Offshore Wind*. Feb. DOE/ GO-102011-3143.

Donnelly, J. 2005. "Price rise and new deep-water technology opened up offshore drilling." *Boston Globe*, Dec. 11. www.boston.com/news/world/articles/2005/ 12/11/price_rise_and_new_deep_water_technology_opened_up_offshore_ drilling.

Economist. 2008a. "Ne plus ultra." *Economist*, Jan. 31. www.economist.com/node/ 10601407.

———. 2008b. "Crude measures." *Economist*, May 29. www.economist.com/node/ 11453151?story_id=11453151.

———. 2011. "Gauging the pressure." *Economist*, April 28. www.economist.com/ node/18621367?story_id=18621367.

Eden, R. J., et al. 1983. *Energy Economics: Growth, Resources and Policies*. Cambridge: Cambridge University Press.

Edison. 2009. "Undergrounding." Edison Electric Institute website. www.eei.org/ourissues/electricitydistribution/Pages/Undergrounding.aspx.

Edwards, B. K. 2003. *The Economics of Hydroelectric Power*. Cheltenham, UK: Edward Elgar Publishing.

Edwards, L. 2010. "World['s] first superconducting DC power transmission system a step closer." Physorg.com, March 8. www.physorg.com/news187251385.html.

EIA (U.S. Energy Information Agency). 2005. *Annual Energy Outlook, 2005*. Feb. DOE/EIA-0383. ftp://ftp.eia.doe.gov/forecasting/0383(2005).pdf.

———. 2010a. *Annual Energy Outlook, 2010*. DOE/EIA-0383(2010). April. www.eia.gov/totalenergy/data/annual/previous.cfm.

———. 2010b. *Annual Energy Review, 2009*. DOE/EIA-0384(2009). Aug. 19. www.eia.gov/totalenergy/data/annual/previous.cfm.

———. 2010c. "Levelized cost of new generation resources in the annual energy outlook, 2011." November. Independent Statistics and Analysis, U.S. Energy Information Administration website. www.eia.gov/oiaf/aeo/electricity_generation.html.

———. 2010d. *Annual Energy Outlook, 2011*. DOE/EIA-0383(2011). Dec. www.eia.gov/forecasts/aeo.

———. 2011. "History of energy consumption in the United States, 1775–2009." *Today in Energy*, Feb. 9, 2011. www.eia.gov/todayinenergy/detail.cfm?id=10.

Encarta. 2005. *Encarta Encyclopedia*. Standard edition 2005. S.v. "Horsepower."

Evans, S. 2012. "Will sun still shine on Germany solar power industry?" BBC, March 12. www.bbc.co.uk/news/business-17344074.

Eveready. 2001. *Eveready Carbon Zinc (ZN/MnO$_2$) Application Manual*. Nov. 6. http://data.energizer.com/PDFs/carbonzinc_appman.pdf.

Fetter, S. 2009. "How long will the world's uranium supplies last?" *Scientific American*, Jan. 26. www.scientificamerican.com/article.cfm?id=how-long-will-global-uranium-deposits-last.

Feynman, R. P., R. B. Leighton, and M. Sands. 1970. *The Feynman Lectures on Physics*. Boston, MA: Addison Wesley Longman.

Flink, J. J. 1990. *The Automobile Age*. Cambridge: MIT Press.

Flynn, A. 2012. "BP stock moves into striking range of recovery." *Wall Street Journal*, March 12.

Fox. 2012. "Freak car accident shuts down major US pipeline until Thursday." Fox Business News, March 5. www.foxbusiness.com/news/2012/03/04/freak-car-accident-shuts-down-major-us-pipeline-until-thursday.

Freese, B. 2003. *Coal: A Human History*. New York: Penguin.

Garvine, R. W., and W. Kempton. 2008. "Assessing the wind field over the continental shelf as a source for electric power." *Journal of Marine Research* 66:751–73.

Genta, G. 1985. *Kinetic Energy Storage: Theory and Practice of Advanced Flywheel Systems.* London: Butterworth.

Gipe, P. 2009. *Wind Energy Basics.* White River Junction, VT: Chelsea Green Publishing.

Goodell, J. 2006. *Big Coal: The Dirty Secret behind America's Energy Future.* New York: Houghton Mifflin.

Grant, P. M., C. Starr, and T. J. Overbye. 2006. "A power grid for the hydrogen economy." *Scientific American,* July.

Green, R., H. Hu, and N. Vasilakos. 2011. "Turning the wind into hydrogen: The long-run impact on electricity prices and generating capacity." *Energy Policy* 39:3992–98.

Greenberg, D. 1979. "A numerical investigation of tidal phenomena in the Bay of Fundy and Gulf of Maine." *Marine Geodesy* 2:161–87.

Greenpeace. 2006. *The Chernobyl Catastrophe: Consequences on Human Health.* Amsterdam: Greenpeace, April. www.greenpeace.org/international/Global/international/planet-2/report/2006/4/chernobylhealthreport.pdf.

Gregory, M. 2006. "India struggles with power theft." BBC News, March 15. http://news.bbc.co.uk/2/hi/business/4802248.stm.

———. 2010. "Why are China's mines so dangerous?" BBC News, Oct. 7. www.bbc.co.uk/news/business-11497070.

Haggett, S. 2009. "High cancer rates confirmed near Canada's oil sands." Reuters report, Feb. 6. www.reuters.com/article/2009/02/06/us-health-oilsands-idUS TRE51568020090206.

Hall, J. 2011. "Failsafe? Inside the Candu safety net." *Toronto Star,* April 1. www.thestar.com/news/insight/article/967796 --failsafe-inside-the-candu-safety-net.

Hamilton, G. 2012. "New policy favors natural gas industry." *Vancouver Sun,* Feb. 4, A6.

Hargreaves, S. 2011. "Saudi Arabia poised to become solar powerhouse." CNN, Nov. 21. http://money.cnn.com/2011/11/21/news/international/saudi_arabia_solar/index.htm.

Harvey, C. E. 1986. *Coal in Appalachia: An Economic Analysis.* Lexington: University Press of Kentucky.

Harvey, F. 2011. "World Bank to limit funding for coal-fired power stations." *Guardian,* April 4. www.guardian.co.uk/environment/2011/apr/04/world-bank-funding-coal-power.

Hasbrouck, R. 1996. "Mitigating lightning hazards." *Science and Technology Review,* May, 4–12.

Hickman, L. 2011. "Fusion power: Is it getting any closer?" *Guardian,* Aug. 23. www.guardian.co.uk/environment/2011/aug/23/fusion-power-is-it-getting-closer.

Hills, R. L. 1996. *Power from Wind: A History of Windmill Technology.* Cambridge: Cambridge University Press.

Hiserodt, E. 2009. "Wind vs. nuclear power: Which is safer?" *New American*. www
.thenewamerican.com/tech/energy/item/7062-wind-vs-nuclear-power-which-
is-safer.

Hoyos, C. 2007. "The new seven sisters: Oil and gas giants dwarf Western rivals."
Financial Times, March. www.ft.com/intl/indepth/7sisters.

Huggins, R. A. 2010. *Energy Storage*. New York: Springer.

Hughes, T. P. 1983. *Networks of Power: Electrification in Western Society, 1880–1930*.
Baltimore: Johns Hopkins University Press.

Hvistendahl, M. 2007. "Coal ash is more radioactive than nuclear waste." *Scientific
American* 112 (Dec.). www.scientificamerican.com/article.cfm?id=coal-ash-is-
more-radioactive-than-nuclear-waste.

IAEA (International Atomic Energy Agency). 2011. "Fukushima Nuclear Accident
Update Log." www.iaea.org/newscenter/news/tsunamiupdate01.html.

IEA (International Energy Agency). 2008. *Key World Energy Statistics*. www.iea.org.

Inman, M. 2010. "Mining the truth on coal supplies." *National Geographic News*,
Sept. 8.

Jackson, D. C. 2005. *Building the Ultimate Dam: John S. Eastwood and the Control of
Water in the West*. Norman: University of Oklahoma Press.

James, P., and N. Thorpe. 1994. *Ancient Inventions*. New York: Ballantine Books.

Johnson, R. C., T. J. Mercier, and M. E. Brownfield. 2011. "Assessment of in-place
oil shale resources of the Green River Formation, Greater Green River Basin in
Wyoming, Colorado, and Utah." U.S. Geological Survey Fact Sheet 2011-3063.
http://pubs.usgs.gov/fs/2011/3063.

Kahya, D., and R. Anderson. 2011. "Carbon dioxide from industry could be buried
offshore." BBC News, Nov. 22. www.bbc.co.uk/news/business-15836202.

Kaku, M., and J. Trainer, eds. 1983. *Nuclear Power: Both Sides*. New York: W.W.
Norton.

Karam, P. A. 2006. "How do fast breeder reactors differ from regular nuclear power
plants?" *Scientific American*, July 17. www.scientificamerican.com/article.cfm?
id=how-do-fast-breeder-react.

Karsten, R. H., et al. 2008. "Assessment of tidal current energy in the Minas Passage,
Bay of Fundy." *Proceedings of the Institute of Mechanical Engineers* 222:493–507.

Keeling, R. F., et al. 2009. "Atmospheric CO_2 records from sites in the SIO air
sampling network." In *Trends: A Compendium of Data on Global Change*. Carbon Dioxide Information Analysis Center, Oak Ridge National Laboratory, U.S.
Department of Energy, Oak Ridge TN. DOI: 10.3334.CDIAC/atg.035.

Khagram, S. 2004. *Dams and Development: Transnational Struggles for Water and
Power*. Ithaca, NY: Cornell University Press.

Kim, V. 2011. "Japan damage could reach $235 billion, World Bank estimates." *Los
Angeles Times*, March 21.

Landels, J. G. 1978. *Engineering in the Ancient World*. London: Constable.

Leeb, S. 2009. *Game Over: How You Can Prosper in a Shattered Economy.* New York: Hachette Book Group.

Lerner, E. J. 2003. "What's wrong with the electric grid?" *Industrial Physicist* (American Institute of Physics), Oct.–Nov. www.tipmagazine.com/tip/INPHFA/vol-9/iss-5/p8.html.

Lienhard, J. 2000. *The Engines of Our Ingenuity: An Engineer Looks at Technology and Culture.* Oxford: Oxford University Press.

Lightfoot, H. D., et al. 2006. "Nuclear fission fuel is inexhaustible." In *EIC Climate Change Technology, 2006.* EIC Climate Change Technology Conference, 2006 IEEE, Ottawa, May. DOI: 10.1109/EICCCC.2006.277268.

Linden, D., and T. B. Reddy, eds. 2002. *Handbook of Batteries.* 3rd ed. New York: McGraw-Hill.

Lockard, D. 1998. *Coal: A Memoir and Critique.* Charlottesville: University Press of Virginia.

Mancuso, J. J., and R. E. Seavoy. 1981. "Precambrian coal or anthraxolite: A source for graphite in high-grade schists and gneisses." *Economic Geology* 76:951–54.

Mankins, J. C. 2002. "A technical overview of the 'Suntower' solar power satellite concept." *Acta Astronautica* 50:369–77.

———, ed. 2011. *Space Solar Power.* Paris: International Academy of Astronautics.

Mason, S. F. 1962. *A History of the Sciences.* New York, Macmillan.

Masters, G. M. 2004. *Renewable and Efficient Electric Power Systems.* Hoboken, NJ: John Wiley and Sons.

Mattick, C., E. Williams, and B. R. Allenby. 2009. "Energy and civilization: A history of energy production and consumption in a global cultural, technological and economic context." In *Proceedings of the 2009 IEEE International Symposium on Sustainable Systems and Technology.* Phoenix, AZ, May. DOI: 10.1109/ISSST .2009.5156766.

Maugeri, L. 2006. *The Age of Oil: The Mythology, History, and Future of the World's Most Controversial Resource.* Westport, CT: Praeger.

McNeil, I., ed. 1990. *An Encyclopaedia of the History of Technology.* New York: Routledge, Chapman and Hall.

Memoli, M., and P. Nicholas. 2010. "BP agrees to $20-billion escrow fund; cancels dividends." *Los Angeles Times,* July 16.

MIT (Massachusetts Institute of Technology). 2011. "The Future of Natural Gas: An Interdisciplinary MIT Study." Cambridge: MIT, June 9.

Morris, N. 2007. *Nuclear Power.* North Mankato, MN: Smart Apple Media.

Mostrous, Y. G., E. H. Gue, and D. F. Dittman. 2010. *Investing in Coal: The World's Workhorse.* Upper Saddle River, NJ: FT Press.

Narisade, K., and D. Schreuder. 2004. *Light Pollution Handbook.* Dordrecht, The Netherlands: Springer.

National Research Council. 1992. *Nuclear Power: Technical and Institutional Options for the Future.* Washington, DC: National Academies Press.

Nersesian, R. L. 2010. *Energy for the 21st Century: A Comprehensive Guide to Conventional and Alternative Sources*. New York: M. E. Sharpe.

Niagara. 2010. International Niagara Board of Control. "One Hundred Fifteenth Semi-Annual Progress Report to the International Joint Commission." www.ijc.org/rel/boards/niagara/115.pdf.

NRC (U.S. Nuclear Regulatory Commission). 2010. *2010–2011 Information Digest*. U.S. Nuclear Regulatory Commission report NUREG-1350, vol. 22. August. pbadupws.nrc.gov/docs/ML1024/ML102460490.pdf.

Overy, R. 2007. *The Times Complete History of the World*. London: Times Books.

Parasiliti, F., and P. Bertoldi. 2003. *Energy Efficiency in Motor Driven Systems*. Heidelberg: Springer.

Parfit, M. 2005. "Future power: Where will the world get its next energy fix?" *National Geographic*, August, 2–31.

Paul-Choudhury, S. 2011. "Japan's megaquake: What we know" *New Scientist*, March 12. www.webcitation.org/5xgjBRle0.

Pendergast, D. R. 1991. "CANDU heavy water reactors and fission fuel conservation." In *Climate Change and Energy Policy*, American Institute of Physics, Proceedings of the International Conference on Global Climate Change: Its Mitigation through Improved Production and Use of Energy, Los Alamos, New Mexico, Oct. 21–24, 1991. www.computare.org/Support documents/Publications/FissionFuel Conservation.htm.

Pérez-Peña, R., and E. Lipton. 2003. "Elusive force may lie at root of blackout." *New York Times*, Sept. 23.

Peters, W. A., et al. 2005. *Sustainable Energy: Choosing among Options*. Cambridge: MIT Press.

Pickard, W. F., N. J. Hansing, and A. Q. Shen. 2009. "Can large-scale advanced adiabatic compressed air energy storage be justified economically in an age of sustainable energy?" *Journal of Renewable and Sustainable Energy* 1:033102.

PSC (Public Service Commission of Wisconsin). 2011. *Underground Electric Transmission Lines*. Public Service Commission of Wisconsin website. http://psc.wi.gov/thelibrary/publications/electric/electric11.pdf.

Pynn, L. 2012. "Critics concerned about safety of 'fracking,' shipping of LNG." *Vancouver Sun*, Feb. 4, A7.

Rajey, A. 1996. *Natural Gas: Production, Processing, Transport*. Paris: Editions Technip.

Ray, R. 2010. "Untapped Potential." *Hydro Review*, Dec. 17. www.renewableenergyworld.com/rea/news/article/2010/12/untapped-potential.

Restivo, S. P. 2005. *Science, Technology and Society: An Encyclopedia*. Oxford: Oxford University Press.

Reuters. 2011. "Pirates hijack U.S.-bound oil tanker off Oman." www.reuters.com/article/2011/02/09/us-oman-supertanker-idUSTRE7182Q220110209.

Reynolds, J. S. 2003. *Stronger than a Hundred Men: A History of the Vertical Water Wheel*. Baltimore: Johns Hopkins University Press.

Riemersma, H., et al. 1981. "Application of superconducting technology to power transformers." *IEEE Transactions on Power Apparatus and Systems* PAS-100: 3398-3407.

Roberts, P. 2004. *The End of Oil*. New York: Houghton Mifflin Harcourt.

Rosen, W. 2010. *The Most Powerful Idea in the World: A Story of Steam, Industry, and Invention*. New York: Random House.

Rudolf, J. C. 2010. "Is 'peak oil' behind us?" Nov. 14. http://green.blogs.nytimes.com/2010/11/14/is-peak-oil-behind-us/?partner=rss&emc=rss.

Sampson, A. 1975. *The Seven Sisters: The Great Oil Companies and the World They Shaped*. New York: Viking.

Scientific American. 2009. *The Future of Nuclear Power*. In-Depth Report. Jan. 26. www.scientificamerican.com/report.cfm?id=nuclear-future.

Simpson, J. 2011. "Amid dire warnings, Canada is missing in action." *Globe and Mail*, Nov. 19, F9.

Smil, V. 2005. *Energy at the Crossroads: Global Perspectives and Uncertainties*. Cambridge: MIT Press.

———. 2010a. *Energy Myth and Realities: Bringing Science to the Energy Policy Debate*. Lenham, MD: AEI Press.

———. 2010b. *Energy Transitions: History, Requirements, Prospects*. Santa Barbara, CA: Praeger.

SoCo. 2009. "Overhead vs. Underground." Southern Colorado Transmission Improvements website. www.socotransmission.com/About/scoping-materials.cfm.

Solomon, S. 2010. *Water: The Epic Struggle for Wealth, Power, and Civilization*. New York: Harper Perennial.

Sovacool, B. 2011. *Contesting the Future of Nuclear Power: A Critical Global Assessment of Atomic Energy*. Singapore: World Scientific.

Speight, T. G. 2007. *Natural Gas: A Basic Handbook*. Houston, TX: Gulf Publishing.

Staniford, S. 2005. "Why oil intensity changed in the US economy." *The Oil Drum*, Oct. 21. www.theoildrum.com/story/2005/10/21/4937/9542.

Stark, R. 2005. *The Victory of Reason*. New York: Random House.

Stec, S., and B. Baraj. 2007. *Energy and Environmental Challenges to Security*. Dordrecht, The Netherlands: Springer.

Stevenson, R. D., and R. J. Wassersug. 1993. "Horsepower from a horse." *Nature* 364:195.

Svoboda, E. 2011. "Is fusion power finally for real?" *Popular Mechanics*, July 21. www.popularmechanics.com/science/energy/next-generation/is-fusion-power-finally-for-real.

Swanson, A. M. 2009. "Photovoltaics power up." *Science* 324:891–92.

Ter-Gazarian, A. 1994. *Energy Storage for Power Systems*. Stevenage, UK: Peter Peregrinus.

Termuehlen, H., and W. Emsperger. 2003. "Importance of coal as a fuel for future power generation." In *Clean and Efficient Coal-Fired Power Plants: Development toward Advanced Technology*. New York: American Society of Mechanical Engineers. DOI: 10.1115/1.801942.

Thurston, R. H. 1878. *A History of the Growth of the Steam Engine*. New York: D. Appleton.

Timoney, K. P. and P. Lee. 2009. "Does the Alberta Tar Sands industry pollute? The scientific evidence." *Open Conservation Biology Journal* 3:65–81.

Tiratsoo, E. N. 1983. *Oil Fields of the World*. Beaconsfield, UK: Scientific Press.

Trench, C. J. 2001. "How pipelines make the oil market work: Their networks, operation, and regulation." Allegro Energy Group, December. www.iatp.org/documents/how-pipelines-make-the-oil-market-work-their-networks-operation-and-regulation.

Tunis, E. 1999. *Colonial Living*. Baltimore: Johns Hopkins University Press.

Tussing, A. R., and B. Tippee. 1995. *The Natural Gas Industry*. Tulsa, OK: Pennwell Books.

Tzimas, E., et al. 2005. *Enhanced Oil Recovery Using Carbon Dioxide in the European Energy System*. European Commission Joint Research Centre Report EUR 21895 EN. Dec. http://science.uwaterloo.ca/mauriced/earth691-duss/CO2_General%20CO2%20Sequestration%20materilas/CO2_EOR_Misciblein%20Europe21895EN.pdf.

UN (United Nations). 2006. *Review of Maritime Transport, 2006*. Geneva, Switzerland: U.N. Conference on Trade and Development. http://unctad.org/en/docs/rmt2006_en.pdf.

University of Delaware. 2007. "Car prototype generates electricity, and cash." *Science Daily*, Dec. 9. www.sciencedaily.com/releases/2007/12/071203133532.htm.

Urbina, I. 2011. Drilling Down series. *New York Times*, miscellaneous issues, Feb.–Dec. www.nytimes.com/interactive/us/DRILLING_DOWN_SERIES.html.

———. 2010. "Despite moratorium, drilling projects move ahead." *New York Times*, May 23.

U.S.–Canada Power System Outage Task Force. 2004. *Final Report on the August 14, 2003, Blackout in the United States and Canada: Causes and Recommendations*. April. https://reports.energy.gov/BlackoutFinal-Web.pdf.

USGAO (U.S. Government Accountability Office). 2005. *Motor Fuels: Understanding the Factors That Influence the Retail Price of Gasoline*. Washington, DC, May. www.gao.gov/new.items/d05525sp.pdf.

USGS (U.S. Geological Survey). 1997. "Radioactive elements in coal and fly ash: Abundance, forms and environmental significance." Fact Sheet FS-163-97. Oct.

Usher, A. P. 1988. *A History of Mechanical Inventions*. New York: Dover.

WCI (World Coal Institute). 2005. *The Coal Resource: A Comprehensive Overview of Coal.* London: World Coal Institute.

Wright, A. 2007. "The Importance of Coal." Presentation at 2007 CAMPUT Conference, Kelowna, BC, April 30. www.camput.org/documents/Wright.pdf.

Yablokov, A., V. B. Nesterenko, and A. V. Nesterenko. 2009. *Chernobyl: Consequences of the Catastrophe for People and the Environment.* Annals of the New York Academy of Sciences, vol. 1181. Boston: Blackwell Publishing, for the New York Academy of Sciences.

Yulsman, T. 2008. "Coal ash is not more radioactive than nuclear waste." *CE Journal,* Dec. 31. Center for Environmental Journalism website. www.cejournal .net/?p=410.

INDEX